Solid State Reactions

MATERIALS SCIENCE AND TECHNOLOGY

EDITORS

ALLEN M. ALPER
GTE Sylvania Inc.
Precision Materials Group
Chemical & Metallurgical Division
Towanda, Pennsylvania

JOHN L. MARGRAVE
Department of Chemistry
Rice University
Houston, Texas

A. S. NOWICK
Henry Krumb School of Mines
Columbia University
New York, New York

A. S. Nowick and B. S. Berry, ANELASTIC RELAXATION IN CRYSTALLINE SOLIDS, 1972

E. A. Nesbitt and J. H. Wernick, RARE EARTH PERMANENT MAGNETS, 1973

W. E. Wallace, RARE EARTH INTERMETALLICS, 1973

J. C. Phillips, BONDS AND BANDS IN SEMICONDUCTORS, 1973

H. Schmalzried, SOLID STATE REACTIONS, 1974

In preparation

J. H. Richardson and R. V. Peterson (editors), SYSTEMATIC MATERIALS ANALYSIS, VOLUMES I, II, AND III

Solid State Reactions

HERMANN SCHMALZRIED

Institut für Theoretische Hüttenkunde und Angewandte Physikalische Chemie
der Technischen Universität Clausthal
Germany

Translated by

A. D. Pelton

Ecole Polytechnique
Departement de génie metallurgique
Montréal 250, Quebec

VERLAG CHEMIE Weinheim/Bergstr. 1974

ACADEMIC PRESS, INC. New York and London
A Subsidiary of Harcourt Brace Jovanovich, Publishers

Revised translation from the German.
Title of the original: "Festkörperreaktionen".

Prof. Dr. Hermann Schmalzried
Institut für Theoretische Hüttenkunde und Angewandte Physikalische Chemie
der Technischen Universität Clausthal
3392 Clausthal-Zellerfeld
Agricolastraße
Germany

This book contains 74 figures and 6 tables

ISBN: 3-527-25560-5 (Verlag Chemie)
0-12-625850-3 (Academic Press)

LIBRARY OF CONGRESS CATALOG CARD NO. 73-17638

Composition: H. Becker, Bad Soden/Ts. – Germany; Printed and bound by Fränkische Gesellschaftsdruckerei Echter Verlag GmbH, Würzburg
Printed in Germany.

Preface

This monograph has been written in order to provide a quantitative understanding of solid state reactions. A large variety of chemical reactions in the solid state exists, many of which are quite complex. Thus, the student as well as the researcher in this field must concentrate on the essentials. It is almost impossible to give a complete survey of the area of solid state reactions in a small volume with a restricted number of pages because this field comprises, among other topics, thermal decomposition, phase transformations, photochemical reactions, corrosion, nucleation and growth, reactions in and with glasses, and sintering. Therefore, I have tried to bring the reader to a modern understanding of the problems by presenting the basic ideas of point defect theory and point defect thermodynamics on one hand, and of phenomenological transport theory on the other. This should enable the reader to understand the literature of the field, to analyze new situations correctly and to formulate new problems quantitatively. I have also preferred to discuss unambiguous limiting cases rather than to describe complicated reactions which cannot be analyzed in a quantitative manner.

There are several books that overlap with the contents of this monograph. However, as far as they are concerned with solid state reactions, they have either been written some time ago or they are quite voluminous and nonselective. Thus, I can only hope that I have supplied chemists and metallurgists with an useful guide to cope with some of their problems.

The English edition of this book has been supplemented in a number of sections. Details have been changed for clarification and a few corrections have been made. I am indebted to co-workers and colleagues for their help and advise. Professor Carl Wagner made a critical reading of the manuscript. Dr. H. Meurer and Ms. A. Kühn helped in proof reading, and Professor Peter Haasen and other colleagues made valuable suggestions. I am especially indebted to Dr. A. D. Pelton who did the cumbersome work of translation.

Clausthal-Zellerfeld, January 1974 H. Schmalzried

Contents

1. Short introduction to the bonding, structure, and imperfections of solids

The goal of this book is to make chemical reactions in the solid state understandable, that is, to describe chemical reactions on the basis of the interactions and the motion of the atomic particles of the systems under consideration. The driving forces for chemical reactions are known if the partial energetic thermodynamic parameters of the reacting system are individually available. However, when we consider the motion of the particles (i. e. the kinetic parameters) many structural questions arise. It is not possible today to make quantitative predictions about the reactivity of a solid solely on the basis of its chemical or physical classification. Still, such classifications are very useful for the development of empirical or semiempirical rules, and for an understanding of the reactivity of various groups of materials. Towards these ends, this introductory chapter is devoted to certain aspects of crystal chemistry which are important to solid state reactions. Special attention will be devoted to those topics which, along with thermodynamics, primarily determine the reactivity of solid phases. These are the bonding, structure, and imperfections of solids [1].

1.1. Classification of crystalline solids

If a certain number of like or unlike atoms are brought together from infinity at a given pressure P and temperature T they will form, by virtue of interactions between the particles, either a gas (in the case of small or zero interactions), a liquid, or a solid. For equilibrium in a system of a given number of particles at a given temperature and pressure, it is necessary that the Gibbs free energy be at a minimum. The spatial configuration of the particles will correspondingly be determined through two parameters: 1. through the energy of interaction which, together with the thermal (kinetic) energy, gives rise to the internal energy U, and 2. through the entropy S which is a quantitative measure of the probability of realizing the thermodynamic state or – less precisely but more intuitively – a measure of the probability of the configuration of the system. Since

$$G = U + PV - TS$$

it can be seen that, at low temperatures, the interaction between particles totally determines the structure, while at higher temperatures the entropy becomes more and more important in influencing the configuration of the particles. Thus, the structure, lattice vibrations, and defects of every crystal in thermodynamic equilibrium result from a compromise between interaction energy and entropy which, together, determine the state of the system through minimization of the Gibbs free energy. The type of interaction between the particles is uniquely given by their atomic number, even though a theoretical understanding of the details of this interaction is still lacking in most cases. In principle, quantum chemistry provides a means of calculating the interactions [2], but the solid systems with which we are dealing are far too complicated to permit much more than qualitative statements to be made. A fundamental task will be to specify the various limiting classes of solids and, if possible, to uniquely characterize them.

The predominant limiting cases of bonding between the atoms of crystals can be classified as follows:

1. heteropolar or ionic bonding
2. covalent bonding
3. metallic bonding
4. van der Waals or molecular bonding
5. hydrogen bonding

Since chemical bonding is restricted to electrons of the outer shell, the difference between the types of bonding can easily be presented schematically by showing the distribution of these electrons and the interaction between them. This is shown in Fig. 1-1 for the main types of bonding. It must be noted however that pure bond types are seldom observed.

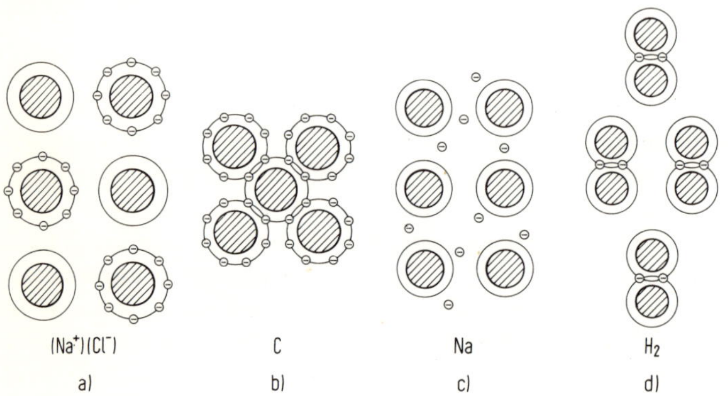

Fig. 1-1. The four main bonding types for solids: (a) ionic bonding (NaCl), (b) covalent bonding (diamond), (c) metallic bonding (Na), (d) molecular bonding (solid H_2). The filled circles represent nuclei and inner electron shells.

To be sure, it is not possible to uniquely predict the crystal structure of a solid phase just on the basis of its bonding characteristics. However, such classifications permit the formulation of rules which allow us to greatly reduce the large number of structures possible for a given substance. All these rules are based on the fact that at $T = 0$ K the coordination polyhedra of a stable structure seek an energy minimum. In addition it is necessary 1. that electroneutrality be preserved, 2. that the directionality of covalent bonds be satisfied, 3. that the repulsion between ions of like charge be kept small, and 4. that, with these stipulations being observed, a closest possible packing be achieved. In the following sections, some characteristics of crystals of the various bonding types will be discussed. Crystals with pure van der Waals bonding will not be treated here, since they will play no role in the framework of this monograph.

1.1.1. Heteropolar crystals

Compounds formed between strongly electropositive (i. e. weakly electronegative) and strongly electronegative atoms comprise the group of so-called ionic crystals. Their internal energy at the absolute zero of temperature is readily understood as the electrostatic lattice energy U_L (if quantum effects are disregarded). This energy can be calculated on the basis of coulombic interaction. This means that the crystal is composed of ions. The electropositive atoms have given up electrons, and the electronegative atoms have gained electrons (see Fig. 1-1). However, in ionic crystals a certain amount of covalent bonding is always found. In SiO_2, for example, this can amount to 50% of the total.

By means of X-ray diffraction or a number of other diffraction methods, the distance between the centers of the ions in a crystal can be measured. These distances can then be used to devise an extensive and consistent system of effective ionic radii for the solid state [2,3]. The numerical values of these so-called ionic radii are slightly dependent upon the coordination number. In Table 1 (see inside back cover) are given, among other data, the effective radii according to Goldschmidt for a coordination number of six.

Up to a coordination number of 57 for the elements in the periodic system, it can be seen that the effective radii increase down a column with increasing atomic number and decrease along a row with increasing positive ionic charge. For higher atomic numbers, this rule must be modified on account of the filling of inner electron shells (lanthanide contraction). In general, it can be stated that at a distance from the nucleus of the order of half the effective radius the electron density has decreased to a few percent [3]. Thus, the intuitive term »ionic radius« is, in the final analysis, misleading, and the term »effective radius« is more appropriate. If the ion possesses an electron shell which is not completely filled, as is the case for transition metal cations, then deviations from spherical symmetry in the effective ionic sphere are indicated by the crystal geometry (crystal field theory) [4]. If, on the other hand, the outer electron shell of the ions is completely filled, then the ionic bonding is non-directional, and the position of the ions in the coordination polyhedron is dependent only upon geometrical requirements and upon the condition of electroneutrality.

If the ions of the crystal are brought closer together than their equilibrium separation distance by the application of a pressure, then rapidly increasing repulsive forces K_r can be evidenced by measurements of the compressibility. These repulsive forces can be described by expressions of the form [19]:

$$K_r = \text{const} \cdot r^{-n} \quad \text{or} \quad K_r = K_r^0 \cdot \exp - (\text{const} \cdot r) \tag{1-1}$$

Together with the coulombic interaction, these forces lead to plots of potential versus interatomic distance r such as that shown schematically in Fig. 1-2.

The exponent n in eq. (1-1) is of the order of ten. For large interatomic separations, the total potential curve follows the coulombic law, and for small separations it follows the repulsive law of eq. (1-1). The equilibrium separation r_{eq} can be calculated from the total potential curve [5, 12].

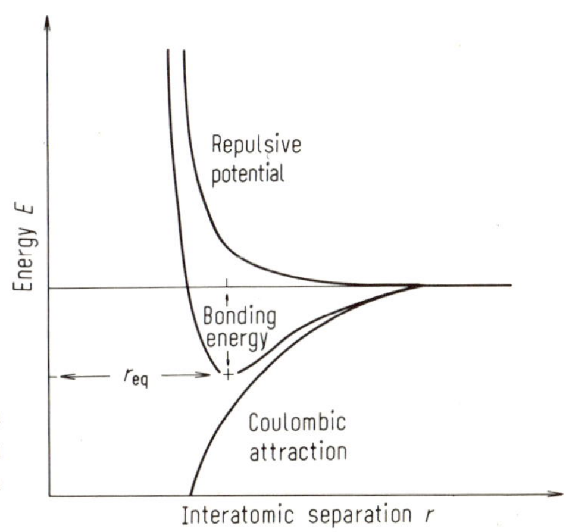

Fig. 1-2. The potential energy of two atoms in a crystal as a function of their separation. When $r = r_{eq}$, then $\partial E/\partial r = 0$. r_{eq} is the equilibrium separation.

By summation over all interionic interactions, the lattice energy U_L is obtained as:

$$U_L = N_0 \alpha_M \frac{2 e_0^2}{d} \left(1 - \frac{1}{n}\right) \tag{1-2}$$

α_M is a dimensionless structure-specific constant known as the Madelung constant, and d is the lattice constant [2,13]. For a lattice with the NaCl structure, $\alpha_M = 1.748$.

Energetically favourable configurations of the ions satisfy Pauling's rules [2,3]. The two most important rules are: 1. The bond strength of an anion in its coordination structure is equal to its charge $\left(6 \cdot \frac{1}{6} = 1 \text{ for the six bonds of a } Cl^- \text{ ion to } Na^+ \text{ ions in NaCl}\right)$. 2. Common edges or surfaces between neighbouring coordination polyhedra diminish the stability of a structure, since like ions can thereby come closer together than they can when the coordination polyhedra share only corners. It is also possible to reach conclusions about the stability of a structure through purely geometrical considerations. Energetically favourable configurations arise when the central ion of a coordination polyhedron touches as many oppositely charged neighbours as possible, thereby avoiding the situation of hanging unsupported or "rattling" inside the polyhedron. Thus, for example, in the case of close anionic packing, tetrahedral coordination of the cations is stable for a radius ratio between $0.225 < r_C/r_A < 0.414$, while octahedral coordination is stable for $0.414 < r_C/r_A < 0.732$ (r_C = cation radius, r_A = anion radius).

In many cases, heteropolar crystals conduct electric current through the motion of ions, and they can be electrolyzed by means of a sufficiently high voltage. Even when, in certain ranges of component activities, electronic partial conductivity predominates in an ionic crystal, its absolute value is always small in comparison with that of normal semiconductors or metals which will be discussed later. One final characteristic property should be mentioned: Ionic crystals absorb strongly in the infrared by virtue of vibrations of the totality of the cations and anions in their sublattices.

1.1.2. Crystals with predominantly covalent bonding

Diamond, a typical example of a homopolar crystal, has a high melting point, and is hard, transparent, and a good insulator. The electron density does not decrease as much along a bond joining two atoms on lattice sites as in the case of ionic crystals [14]. The individual atoms of homopolar crystals possess only partially filled orbitals, and these orbitals can therefore be mutually filled up by the electrons of neighbouring atoms in the crystal. Expressed somewhat differently, this means that a fraction of the electrons of the outer shell belong to more than one atom. In this case we speak of electron pair bonding (see Fig. 1-1). When carbon atoms are brought together to form a diamond crystal, one of the 2s-electrons is elevated into the third unoccupied 3p-orbital. The four orbitals, which are now half-filled, group themselves together into so-called (sp^3)-hybrids. These orbitals then form the well-known tetrahedrally oriented electron pair bonds of diamond. Compared to a closest packing of spheres, the diamond structure is relatively open. This shows that for covalent bonding the preferred situation is to achieve the greatest possible overlap of the orbitals of neighbouring atoms. The energy minimum is essentially no longer determined by the packing density.

For other elements, this type of reasoning leads easily to the so-called (8-n)-rule [2]. If n is the number of electrons in the outer shell, then the number of nearest neighbours for the case of electron pair bonding is equal to (8-n). Examples of tetrahedral four-fold coordination in which all bonds are primarily covalent are crystals of carbon, silicon, germanium, and gray tin. Phosphorus and arsenic are examples of crystals with three-fold coordination. Arsenic has an hexagonal layer structure with only weak bonding between the hexagonal layers. Sulfur, selenium, and tellurium, which should exhibit two-fold coordination according to the rule, form rings and chains which fulfill this requirement.

Crystals with predominantly covalent bonding are typical semiconductors (Ge, Si, InSb). Electrical conduction arises through the freeing of localized bonding electrons. This can occur because of thermal vibrations of the crystal or because of incident light. The free electrons belong to the crystal as a whole. Taken together, these so-called conduction electrons form a sort of electron gas which is responsible for the electrical conductivity. The missing electrons in the electron pair bonds can be replaced from neighbouring bonds. Under the influence of an electrical field, electrical conductance can arise due to the directional migration of these "electron holes", just as for the case of the migration of conduction electrons. In this connection we speak of hole conduction in the valence band [6]. The band picture can be understood as follows (see Fig. 1-3):

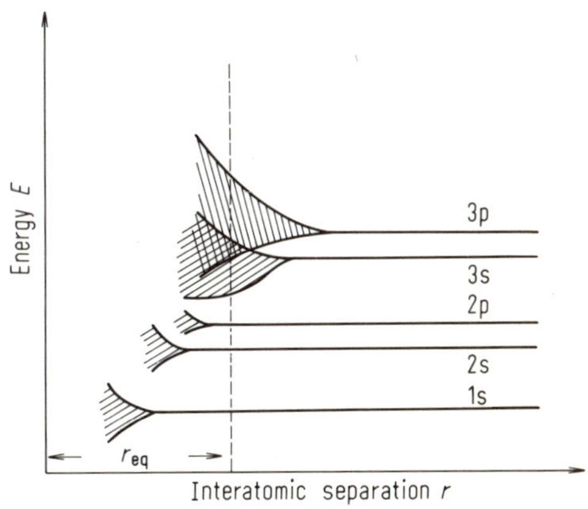

Fig. 1-3. Schematic representation of the electron energy for Mg as a function of the interatomic separation. r_{eq} = equilibrium distance.

If the atoms of a semiconductor are brought together to form a crystal, then interaction between the electrons of the outer shell commences. An atomic electronic state is thereby split up in proportion to its interaction with all the other atoms of the crystal. Each resultant state can take up two electrons with opposite spin. The electrons now no longer belong to the single atom, but rather to the entire crystal as a whole. Thus, each atomic electronic state gives rise to a band of states in the crystal. The completely filled band of highest energy is called the valence band. Electrons can be elevated from the valence band of a covalent crystal into the next highest band, the so-called conduction band, if enough energy is supplied (as, for example, thermal energy). This has already been discussed above.

1.1.3. Crystals with predominantly metallic bonding

Metals are formed by atoms which possess a smaller number of outer electrons than nearest neighbours. Alkali metals with one outer electron or alkaline earth metals with two outer electrons are typical examples. Metals crystallize preferentially in close-packed configurations with high coordination numbers. In the metal, the outer electrons behave essentially as if they

were quasi-free, since the upper electronic energy band is not completely filled and thus the electrons in this conduction band can take up kinetic energy. We speak of a "degenerate" electron gas which exists in the potential field of all the cations. The expression "degenerate" means that the electrons in the metal do not behave like an ordinary gas which obeys Boltzmann statistics, but rather, because of the finite number of electrons in the energy states of a band, they obey another type of statistics, the so-called "Fermi statistics".
The influence of the periodic potential of the crystal upon the motion of the electrons under the action of a force can be taken into account by considering the electrons to possess a so-called effective mass m^* which is different from their rest mass m. The concept of the electron gas explains the most important metallic properties of solid materials: metallic bonding, good thermal and electrical conductivity, high absorptivity or reflectivity of visible light, and last but not least, good plastic deformability [7].

The concept of quasi-free conduction electrons implies that their scattering by the ion core potential in the solid is rather weak. From here the modern theory of the "pseudopotential" has been developed. This theory shows that it is possible to reproduce the scattering of electron waves by replacing the deep potential at each site of the ionic core by a very much weaker effective potential, the pseudopotential. Thus the total pseudopotential in the metal or the semiconductor, which the conduction electrons feel, is fairly uniform, and the replacement of the real potential by the pseudopotential is a perfectly rigorous procedure. Furthermore, the fact that the total pseudopotential is fairly flat means that one can apply perturbation theory in order to calculate electron energies, cohesion, optical properties, etc. [20].

Different metals which crystallize in one and the same structure and whose lattice constants do not differ much from each other exhibit extended ranges of mutual solid solubility. At lower temperatures the different types of atoms can become ordered in simple ratios, while at higher temperatures they are distributed randomly over the lattice sites. This ordering at lower temperatures gives rise to so-called superstructures (e.g. Cu_3Au).

Metals frequently form intermetallic compounds, particularly when the individual pure metals have different structures. Intermetallic compounds exhibit a more (Cu-Mg) or less (Cu-Zn) narrow range of homogeneity. Metallic bonding is the predominant type of bonding in many intermetallic compounds. However, a certain fraction of the other types of bonding must also be taken into consideration. This partially non-metallic character of the bonds, as well as the different sizes of the atoms, is responsible for the large variety of crystal structures of intermetallic compounds. Only some of the most important groups will be mentioned here.

Crystal structures of the so-called Hume-Rothery-phases are connected with certain average numbers z of the valence electrons of the alloy [8,15] (β-brass, CuZn, $z = 3/2$; γ-brass, Cu_5Zn_8, $z = 21/13$). When counting up the valence electrons it must be noted that for the metals Fe, Co, Ni, and the platinum group the number of valence electrons is not given by position in the periodic table, but rather, the valencies must be set equal to zero.

For the so-called Laves-phases [8, 16] of the general form AB_2 ($MgCu_2$, KNa_2) the quotient of the atomic radii $r_A/r_B \simeq 1.25$ is evidently of prime importance in determining the structure, although the valence electron concentration and the portion of non-metallic bonding also affect the finer features of the structure.

For the so-called intercalation (interstitial) phases [8] (carbides, nitrides, hydrides) the most important role is played by the atomic size. If the non-metallic atom B is small enough ($r_B/r_A < 0.6$) then it can be introduced into the interstices of the structure of the metal A – as for example into the center of the cubic elementary cell of a cubic face-centered lattice of metal A.

Any deviation of an intermetallic compound from the strict stoichiometric composition always means that there is a disturbance in the ideal order of the crystal lattice. That is to say, defects are present. Suppose that the isothermal activity of metal A (i.e. the relative vapour pressure compared to the pure material) is measured in a binary system A-B in which intermetallic compounds with finite ranges of homogeneity exist. Then, for an arbitrarily chosen phase diagram, a plot of activity versus composition as shown schematically in Fig. 1-4 results.

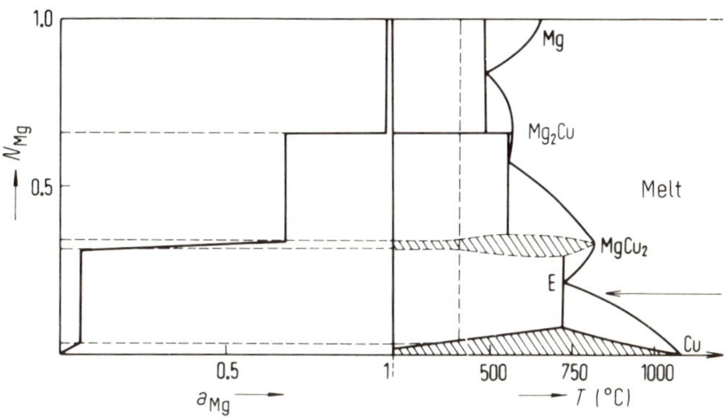

Fig. 1-4. Schematic phase diagram Cu-Mg and the corresponding plot of activity of Mg at 400 °C; N_{Mg} = mole fraction of Mg; E = eutectic.

From the relationship

$$\mu_A = \mu_A^0 + RT \ln a_A \tag{1-3}$$

it can be seen that the activity a_A of A is a measure of the partial molar free energy μ_A (or the chemical potential) of component A. The more stable an intermetallic compound is, then, in general, the steeper is the decrease in the activity of A with increasing mole fraction of B over the range of homogeneity. The activity-composition curve is determined by the atomic disorder. An exact treatment of this subject is provided by defect thermodynamics which will be discussed in chapter 4.

1.2. Defect structure and microstructure of solids

Strictly speaking, the concept of crystallographic structure is applicable only to crystals which are infinitely extended in three dimensions. In reality, we have at our disposal at best relatively defect free single crystals of a limited size, and the periodicity of the crystal lattice ends at the surface. In addition, disturbances of the strict periodicity of the crystal arise because of point defects (vacancies, interstitial atoms, substitutional atoms), and also because of one-dimensional crystal defects (dislocations), two-dimensional defects (low- and high-angle grain boundaries), and in certain cases three-dimensional defects (inclusions, pores). Point defects can achieve thermodynamic equilibrium if given a long enough time, provided that the temperature is not too low. However, the concentrations of all other defects depend upon the way in which the crystal was formed and subsequently handled. These defects are frozen in, and are not in thermodynamic equilibrium. Point defects and dislocations, which will be discussed later in much detail, are structural defects, while all other irregularities in the crystal are microscopic irregularities.

Some methods of growing single crystals may be briefly mentioned [9]: 1. epitaxial deposition of the solid from the gas phase, 2. oriented crystallization from supersaturated solutions, 3. controlled solidification from melts. To this final category belong the most common techniques such as methods of drawing from the melt with a seed crystal (Czochralski, Kyropoulos), the zone melting method (Bridgman-Stockbarger), or the flame melting technique (Verneuil). Recrystallization of polycrystalline solids can likewise give serviceable single crystals.

It is almost impossible to avoid introducing a large number of the crystal defects mentioned above during the preparation of material with a high melting point. The reasons for this are firstly that the surrounding gas phase will almost always contain foreign matter which is then introduced into the crystal, and secondly, that steep temperature gradients and the associated mechanical stresses lead to local plastic flow and to the formation of dislocations. During the preparation of single crystals of an alloy or of an intermetallic compound with a finite range of homogeneity, compositional inhomogeneities will generally appear, and there will be concentration gradients of the components in the crystal. By means of an homogenizing anneal these defects can become evened out through diffusional processes.

By far, most solid materials are not found as single crystals, but rather they consist of an aggregate of tiny crystals, and are classified as polycrystalline materials. For single-phase polycrystalline solids, the grain boundaries between the individual crystallites, as well as the structural defects of the tiny single crystals, must be taken into account as imperfections. Single phase materials are said to have a homogeneous microstructure. If the crystallites should grow (that is, if the grain boundaries should migrate) then the total surface energy will decrease. The dislocation density, and therefore the total dislocation energy, will also decrease. Thus, by means of annealing, a state of lower free energy can be achieved by virtue of the increased grain size.

In general, the densities of polycrystalline metals are comparable to those of single crystals. On the other hand, when dealing with non-metallic inorganic compounds with high melting points, it is often difficult to prepare dense material which is free of pores. The most important reasons for this are the relatively high solubility of gases in thc liquids as compared to the solids, and also the slow rate of sintering, especially when the initial material is coarse-grained. Grain boundaries are sites where impurities do preferentially collect. Pores can be open to the surface and connected to one another, or they can be isolated. For transport processes occurring in the solid they give rise to a resistance different from that of the matrix. The determination of the shape, size, and distribution of pores is one of the most difficult problems encountered in the characterization of the microstructure of a solid, particularly for the case of a ceramic material [17].

If, because of the method of preparation, the crystallographic orientations of the individual grains of the polycrystalline aggregate are not randomly distributed, but rather exhibit a preferential orientation, then we speak of a texture to the material. This anisotropy can also give rise to an anisotropy of the properties of polycrystalline materials.

Multiphase solids are, of necessity, polycrystalline. In such cases we speak of a heterogeneous microstructure. The formation of this microstructure depends essentially upon the phase diagram, upon the thermal history of the sample during preparation, and upon the transport properties for mass and heat transport of the phases which are already present as well as of those phases which are forming during preparation [10,11]. Nucleation likewise often plays an important role in determining the microstructure. The microstructure of a solid is thus only completely described when the number of phases, their corresponding proportions, and the characteristics of each individual phase such as grain size distribution, shape, and orientation, and porosity are given. All these characteristics influence the physical properties as well as the chemical reactivity of a solid. Therefore, one of the basic problems of solid state chemistry is the fundamental characterization of a material.

As an important example of the formation of a microstructure, let us consider eutectic solidification [18]. During the cooling of an alloy melt into the two-phase solid region below the eutectic line as shown by an arrow in Fig. 1-4, a solid phase first begins to crystallize out with a composition different from that of the melt. This causes the composition of the melt and the solidifying crystals to change until the eutectic point E is reached. At this point the melt is at the lowest possible temperature at which it can coexist with the solid phases. This we can easily read off the phase diagram. From now on the remainder of the melt solidifies at the eutectic temperature with the simultaneous separation of the solid phases. The formation of the microstructure is thus regulated by diffusional processes, by the heat of solidification, by heat transport, and in certain cases by surface tension inasmuch as this affects nucleation and the stability of the advancing phase boundaries.

1.3. Literature

General Literature:

[1] J. A. Hedvall, Einführung in die Festkörperchemie, Friedr. Vieweg und Sohn, Braunschweig 1952.
[2] L. Pauling, The nature of the chemical bond, 3rd ed., Cornell University Press, N.Y. 1960.
[3] J. Zemann, Kristallchemie, Walter de Gruyter und Co., Berlin 1966, Sammlung Göschen Bd. 1220/1220a.
[4] H. L. Schläfer and G. Gliemann, Einführung in die Ligandenfeldtheorie, Akademische Verlagsgesellschaft, Frankfurt 1967.
[5] F. Seitz, The Modern Theory of Solids, 1st ed., McGraw-Hill Book Comp., Inc., New York 1940.
[6] E. Spenke, Elektronische Halbleiter, 2nd ed., Springer-Verlag, Berlin 1965.
[7] G. E. R. Schulze, Metallphysik, Akademie-Verlag, Berlin 1967.
[8] C. S. Barret, Structure of Metals, 2nd ed., McGraw-Hill Book Comp., Inc., New York 1952.
[9] A. Smakula, Einkristalle, Springer-Verlag, Berlin 1962.
[10] P. G. Shewmon, Transformations in Metals, McGraw-Hill Book Comp., Inc., New York 1969.
[11] H. Salmang and H. Scholze, Die physikalischen und chemischen Grundlagen der Keramik, 5th ed., Springer-Verlag, Berlin 1968.

Special Literature:

[12] M. Born and J. E. Mayer, Z. Physik *75*, 1 (1932).
[13] E. Madelung, Physik. Z. *19*, 524 (1918).
[14] E. Wölfel et al., Z. phys. Chem. NF *4*, 36 (1955), Z. Elektrochem. *63*, 891 (1959).
[15] W. Hume-Rothery, The Structure of Metals and Alloys, Institute of Metals, 8th ed., London 1962.
[16] F. Laves and H. Witte, Metallwirtschaft *14*, 645 (1935).
[17] L. Zagar, Silicates Industriels *30*, 487 (1965).
[18] W. A. Tiller in R. W. Cahn, Physical Metallurgy, North-Holland Publ. Co., Amsterdam 1965.
[19] W. H. Lee and M. F. C. Ladd in H. Reiss, Progress in Solid State Chemistry, Pergamon Press, Oxford 1967, Vol. 3, 265.
[20] V. Heine in F. Seitz and D. Turnbull, Solid State Physics, Academic Press, New York 1970, Vol. 24.

2. Short introduction to solid state reactions

2.1. General remarks

In the first chapter, some concepts of importance to solid state chemistry were explained. In the present chapter, the essential questions which must be answered in connection with solid state reactions will be outlined by means of a few examples. In later chapters, these questions will be treated in more detail, and, wherever possible, quantitatively.

A solid state chemical reaction in the classical sense occurs when local transport of matter is observed in crystalline phases. This definition does not mean that gaseous or liquid phases may not take part in solid state reactions. However, it does mean that the reaction product occurs as a solid phase. Thus, the tarnishing of metals during dry or wet oxidation is considered to be a solid state reaction.

If the reactants are brought together at constant pressure and temperature in a closed system, then the reaction will take place spontaneously if the Gibbs free energy of the system is thereby decreased. The interactions between the atoms in a crystal are relatively large compared with the interactions between particles in other states of matter. Therefore, during the course of a solid state reaction, the electronic configuration of the atoms in the lattice can be appreciably altered. This is connected with local changes of the chemical potentials (i.e. the partial molar free energies μ_i) of the individual components i of the crystal. Therefore, in a narrower sense, solid state chemical reactions are characterized by the diffusion of atoms of type i, or of ions of type i with valence (charge) z_i in chemical (grad μ_i) or electrochemical (grad η_i) potential gradients. The electrochemical potential gradient is defined as:

$$\text{grad}\ \eta_i = \text{grad}\ (\mu_i + z_i F\phi)$$

F is the Faraday constant (96,500 amp·sec). In ionic systems, the electrochemical potential η_i must be used, since in non-metals an electric field $\mathfrak{E} = -\ \text{grad}\ \phi$ generally acts on the atoms moving in an electrical potential ϕ during the reaction, and so an electrical force is also active. Thus, the local changes of the partial molar free energies of the various types of particles are the driving forces for chemical reactions in the solid state. The rate of diffusion is proportional to the driving force. The factors of proportionality are the so-called transport coefficients (the rate constants) of the particular system. A large part of this book is devoted to the atomistic interpretation of these transport coefficients.

The gradients of the chemical or electrochemical potential are not the only possible driving forces for a solid state reaction. For example, an initially homogeneous solid solution can become demixed under the action of a relative temperature gradient $1/T \cdot \text{grad}\ T$. This phenomenon is known as thermal diffusion. Furthermore, we speak of electrolysis in the solid state when ions in ionic crystals migrate solely through the action of an electrical potential gradient grad ϕ. As yet another example, we may consider the sintering process during which solids assume the state of minimum surface area, since this is the state of minimum surface free energy. This occurs by the diffusion of atoms from areas of high surface curvature to areas of lower curvature. This is also a solid state reaction.

A question to be considered is whether the local fluxes of matter in multicomponent systems are coupled with each other so that, for example, a flux of component i induces a flux of component j. The simplest case would be that in which there was no coupling between the fluxes of the various particles during solid state diffusion except that arising from the condition of local electrical neutrality. Then, during the chemical reaction, the local particle flux j_i would be proportional only to the local chemical or electrochemical potential gradient grad μ_i or grad η_i. The range of applicability of this linear dependence is similar in extent to that of the corresponding linear regions for heat transport (Fourier's law) or for electrical transport (Ohm's law). Deviations will be discussed later for individual cases.

The transport of matter in the solid state, and thus also the reactivity of solids, are dependent upon the mobility of the individual particles in the lattice. An ideal ordered crystal can only be moved as a whole, and motion of the individual particles from their lattice sites cannot occur. Therefore, every case of mass transport in solid phases is directly dependent upon deviations from ideal crystalline order. The rate constants (i.e. the transport coefficients) in crystalline phases are directly related to the atomic disorder and usually one finds that the higher the atomic disorder, the higher is the corresponding transport coefficient. In order to permit a detailed look at all the facts, it is advantageous to divide solid state reactions into the following groups in a manner analogous to the familiar classification of general chemical kinetics:

1. Homogeneous reactions
2. Reactions in single phase inhomogeneous systems
3. Heterogeneous reactions

The characteristic of heterogeneous reactions is the occurrence of phase boundaries across which mass transport takes place.

Chemical reactions between solid crystalline materials are in general exothermic. The reason for this is the high degree of order of crystalline phases, which means that the entropy difference between these phases is relatively small. Nevertheless, most solid state chemical reactions can easily be carried out isothermally, since, by virtue of the low reaction rates, the quantity of heat given off per unit time can easily be conducted away. Exceptions from this rule will be dealt with separately at appropriate places.

After these introductory remarks, some typical solid state reactions may be qualitatively discussed.

2.1.1. Homogeneous reactions in the solid state

Let us consider, as an example, oxide or sulphide phases of the general formula AB_2O_4 or AB_2S_4 which crystallize in the spinel structure [8]. In this structure two limiting cases can be distinguished: 1. In the normal spinel the divalent A cations are found on regular lattice sites with tetrahedral symmetry, and the trivalent B cations are on regular lattice sites with octahedral symmetry. 2. In the inverse spinel structure, half of the B cations are found on regular lattice sites with tetrahedral symmetry and the other half of the B cations together with the A cations are on regular lattice sites of octahedral symmetry. Frequently however, the distribution of the A and B cations on the regular lattice sites does not conform to one of these limiting cases. Rather, the A and B cations are distributed over both sublattices, and their equilibrium distribution is clearly temperature dependent [9]. If the temperature of a spinel crystal such as $NiAl_2O_4$, which is at equilibrium, is changed, then a local redistribution of the Ni^{2+} and Al^{3+} ions takes place. For the distribution of the cations between the tetrahedral (index A) and octahedral (index B) sublattices the following reaction equation can be formulated:

$$Ni_A^{2+} + Al_B^{3+} = Al_A^{3+} + Ni_B^{2+}; \qquad \Delta G_1^0 \tag{2-1}$$

Reaction (2–1) proceeds as a homogeneous reaction until the crystal has achieved a new minimum of the total free energy as given by the temperature dependent equilibrium constant of eq. (2–1):

$$K_1\,(T) = \exp\,(-\Delta G_1^0/R\,T) \tag{2-2}$$

By means of diffusional steps in the interstices, the ions of the one sublattice seek out the vacancies in the other sublattice, and a reversal of positions occurs. In this way, equilibration is achieved.

A further example from the area of homogeneous solid state reactions is the formation of so-called Frenkel defects in silver bromide. As will be discussed later in detail, silver ions to a small extent leave their regular lattice sites because of thermal excitation and force themselves into interstices, leaving behind silver ion vacancies in the regular silver ion lattice of the silver bromide [1]. Again, the equilibrium concentrations of these defects are temperature dependent through the standard free energy of reaction, so that a shift in temperature leads to a new distribution of the vacancies and interstitial ions by means of a homogeneous reaction. As can be seen, when dealing with homogeneous reactions in solids we are always concerned with defect reactions. Therefore, defect reactions are very important in the theory of solid state reactions. This is true firstly because during a reaction in inhomogeneous or heterogeneous systems the defect equilibria are often maintained despite local compositional changes in the crystal which occur due to local fluxes of the components (local thermodynamic equilibrium), and secondly because the defects influence many physical and reactive properties of the solid, and in a state in which the defects are frozen into the crystal their concentrations are largely dependent on the rates of the defect reactions during the freezing-in period.

2.1.2. Reactions in a single phase inhomogeneous system

To this class of solid state reactions belong, in general, the familiar diffusional processes in single phase systems with concentration gradients [2]. For example, if iron is packed in graphite powder at 950 °C the carbon atoms diffuse in the interstices of the face-centered cubic austenite down their concentration gradient from the surface into the interior of the iron. Diffusional processes of carbon in iron, which are of decided importance in surface hardening, belong to the class of solid state reactions of inhomogeneous systems. Similarly, semiconductors can be doped (i.e. they can be supplied with low concentrations of foreign elements) in order to control their electrical conduction properties. The surface concentration of the diffusing element is always given by the coexistence of the material being doped with another phase. In this section, however, we are only concerned with bulk diffusional processes.

The simplest limiting case occurs when the pure reactants are completely miscible with each other. Examples of this are chemical diffusional processes which proceed when two crystals of silver and gold, MgO and NiO, or NaBr and AgBr are placed next to each other and are annealed at temperatures where diffusion occurs. Equilibrium is attained when the system becomes homogeneous through diffusion. Diffusion of the various components of the inhomogeneous system takes place by means of defects. In most binary metallic systems, especially when they are close-packed (as, for example, in the system silver-gold), the transport of atoms occurs via vacancies. In ionic crystals the determination of the defect center which is responsible for the mass transport can be a very difficult problem in many cases. Viewed from the phenomenological standpoint, the problem of solid state reactions in single phase inhomogeneous systems lies in the formulation of the differential equations of diffusion and in their solution under given experimental initial and boundary conditions. The transport coefficients in these equations, and especially their concentration dependences, are either derived from defect thermodynamics, or, in the case of experimentally determinable transport coefficients they may be interpreted atomistically, again with the help of defect thermodynamics.

2.1.3. Heterogeneous reactions

If two substances react with one another to form one or more product phases which are separated from the reactants and from each other by phase boundaries, then a heterogeneous solid state reaction is said to occur [3]. A basically analogous case is the formation of an oxide coating on a metal [4]. In purely metallic systems, the simplest product phases are binary, as in the reaction equation

$$nA + mB = A_nB_m$$

An example is the reaction $Al + Sb = AlSb$. In solid heteropolar systems the simplest case is the formation of ternary compounds as in the reaction $CaO + SiO_2 = CaSiO_3$. The reaction product separates the reactants from one another, and the reaction proceeds by diffusion of the participating components through the reaction product. For very low solubilities of the reactants in the reaction product (i. e. for a product with a very narrow range of homogeneity) the particle fluxes are locally constant, and as long as local thermodynamic equilibrium is maintained at the phase boundaries, a parabolic growth law results. We shall return to this matter later. As an example, a spinel formation reaction as shown in Fig. 2-1 will be described. From a single crystal of NiO and a single crystal of Al_2O_3, a partially monocrystalline spinel $NiAl_2O_4$ is formed. In this case the fluxes of the components in the reaction product, which are responsible for the advancement of the reaction, are fluxes of charged particles. Therefore, in order to preserve local electrical neutrality, the fluxes of the different ions must always be coupled with each other. Consequently, the following combinations are possible: either

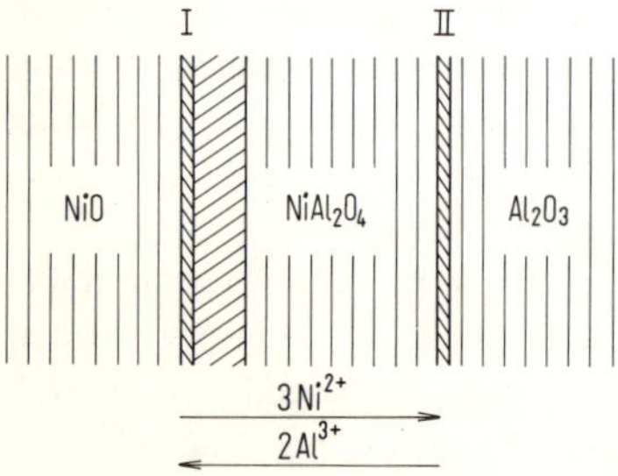

Fig. 2-1. Schematic diagram showing the mechanism of formation of $NiAl_2O_4$ from NiO and Al_2O_3. The reactions at the phase boundaries are
(I) $2Al^{3+} + 4NiO = NiAl_2O_4 + 3Ni^{2+}$
(II) $3Ni^{2+} + 4Al_2O_3 = 3NiAl_2O_4 + 2Al^{3+}$
||| means monocrystalline
/// means polycrystalline

oppositely charged ions flow in the same direction, or ions with like charges flow in opposite directions through the reaction product. These simple considerations give the possible limiting cases of a solid state reaction between ionic crystals. For metallic systems, the condition of electroneutrality obviously can not be expressed in this manner. Although, for certain transport processes in metallic systems, it is also advantageous to speak of separate motion of the ions and electrons (electrotransport [13,6]), it is generally more suitable to assume that atoms migrate. The situation, then, is fundamentally altered: while for the formation of ionic crystals it is the slower partner that essentially determines the reaction rate, in metallic systems it is the faster partner that is responsible for the advancement of the reaction. Examples of the formation of intermetallic compounds are the formation of β-Al_3Mg_2 and δ-Al_2Mg_3 in the diffusion couple aluminum-magnesium, or the formation of various brass phases in the diffusion couple copper-zinc [5].

It is quite possible that not all of the equilibrium phases which appear on the phase diagram will actually appear between the reactants during a heterogeneous solid state reaction. This arises either because of nucleation difficulties or because the particle fluxes are blocked when crossing certain phase boundaries.

2.2. Mass transport in electrical potential gradients

If a homogeneous solid is brought into an electric field by means of two electrodes and an applied voltage, then particle currents will arise corresponding to the local electrical potential gradient and to the transport coefficients of the various charged structural elements of the crystal. Depending upon the type of electrodes and the magnitude of the applied voltage, various effects can occur. First of all, proportionality between the particle currents and the driving force (i.e. the electrical potential gradient) may be assumed. Electrolysis in the proper sense can be said to occur when electrodes which are not reversible to the ions Me^+ and X^- are placed upon an ionic crystal MeX, and when the applied direct voltage E exceeds the decomposition voltage $E_D = \Delta G^0_{MeX}/F$. $E_D F$ is the electrical work equivalent of the free energy of formation ΔG^0_{MeX} for the formation of MeX from the elements Me and $\frac{1}{2}X_2$ (g). Thus, for example, silver bromide can be decomposed at 277 °C by means of platinum electrodes and an applied voltage in excess of 0.87 volts. Silver is deposited at the cathode, and bromine gas is evolved at the anode.

If the applied direct voltage is less than the decomposition voltage, then various situations can arise depending upon the type of electrode. Some of these cases may be listed [10]:

1. The electrodes are reversible to electronic charge carriers and to one type of ion. Then, by measurement of the voltage and current, the total conductivity can be determined.

2. The electrodes are only reversible to electronic charge carriers and are inert to all others. The activities of the components of the crystal are fixed – for instance via the gas phase. With this configuration, the electronic partial conductivity can be measured as a function of the component activities.

3. Essentially the same information can be obtained when both electrodes are reversible to electrons but when the activity of the metallic (non-metallic) component is fixed at the cathode (anode) and an inert gas atmosphere is used. In this case, the migration of ions at steady state

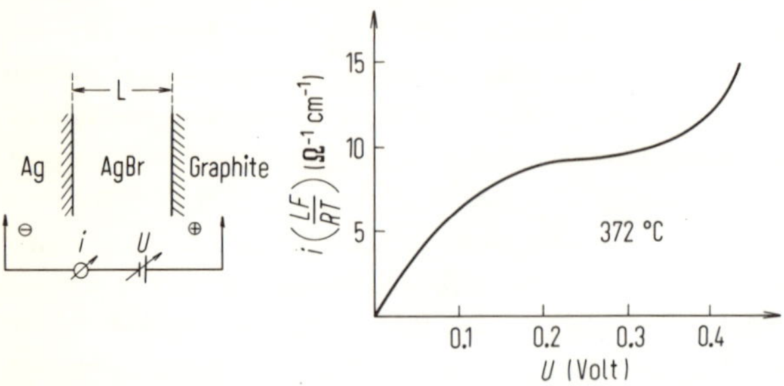

Fig. 2-2. Normalized current density i as a function of the applied voltage U for silver bromide with Ag as cathode and graphite as anode [14]. The ionic current is blocked, and the electronic current is measured. L is the sample length.

will also be prevented by polarization. In Fig. 2-2 are shown the construction of the cell Ag/AgBr/graphite and also the results of measurements using this cell.
Polarization measurements of this type are useful for the determination of transference numbers when the electronic partial conductivity σ_{el} is small compared to the ionic partial conductivity σ_{ion}. From the form of the curve of σ_{el} as a function of the applied voltage, information can be obtained regarding the nature of the electronic charge carriers [14].

4. On the same principle, by the use of solid electrodes of the second kind, the electronic part of the current in a mixed conductor MeX can be blocked, and the ionic current can be measured.

A few final comments should be made. Every phase boundary reaction, and thus every transfer of electrons and ions from the electrode to the crystal MeX and back again, requires a finite driving force. Therefore, electrodes can never be fully reversible. Furthermore in the preceding discussion only ionic crystals have been considered. Here the transport experiments are most easily and clearly interpreted. However, we must also deal with electrolytic transport in semiconductors and metals, although this may be a very small effect. For further details one should refer to the special literature [6].

2.3. Thermal diffusion

If a homogeneous Pb-Sn alloy is introduced into a temperature gradient, then an enrichment of Sn at the hot end and of Pb at the cold end will take place. A still more impressive example of the transport of particles in a lattice in a temperature gradient is observed when Ag_2S is placed in a temperature gradient above 200 °C in an atmosphere of sulphur vapour. Ag_2S disappears from the hot end of the sample, and the same amount of Ag_2S reappears at the colder end. During this process an equal number of equivalents of Ag^+ ions and of electrons migrate in the Ag_2S, while the sulphur is transported via the gas phase.

If the transport of sulphur via the gas phase is prevented, then the transport of silver in the Ag_2S also ceases under steady state conditions. A concentration gradient (i.e. an activity gradient) is built up, and the well-known Soret effect [11] is observed. An even simpler situation arises if the chemical potential of the silver in the sample is maintained constant and equal, for example, to the standard potential μ^0_{Ag} of pure silver. The flux of silver is then proportional to the temperature gradient dT/dx, to the mobility of the silver ions, and to the sum of the heat of solution $\bar{H}_{Ag} - H^0_{Ag}$ and the heat of transfer Q_{Ag} of the silver in the Ag_2S. In view of the small temperature coefficients of the partial conductivities of Ag^+ ions and of electrons, it can be assumed that Q_{Ag} is negligible compared to $\bar{H}_{Ag} - H^0_{Ag}$ [15].

2.4. Sintering

The hypothetical experiment shown in Fig. 2-3 illustrates the basic problem of those solid state reactions known as sintering processes.

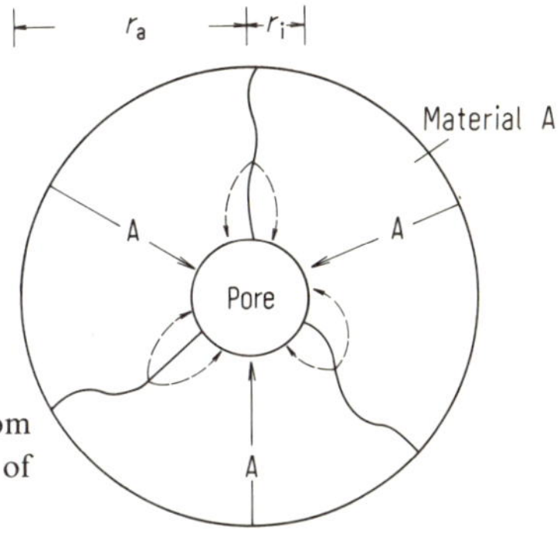

Fig. 2-3. Basic process of sintering: Material A is transported from outer surfaces and grain boundaries to the pore; r_i is the radius of the pore; r_a is the radius of the specimen.

The case which is illustrated applies especially to the advanced stages of sintering when the porosity is limited and the pores are already closed up. Since the curvatures of the inner and outer surfaces of the material A are different, a chemical potential difference exists, which is derived from the surface free energy γ_0. This can be calculated from the Gibbs-Thomson equation under the assumption that $r_i \ll r_a$.

$$\Delta \mu_A = 2\gamma_0 V_m (1/r_i - 1/r_a) \simeq \frac{2\gamma_0 V_m}{r_i} \quad (2\text{-}3)$$

Accordingly, there will be a particle flux from the outer surface to the pore corresponding to the particle mobility. This will fill up the pore. It should be stressed once again that the cause of the solid state reaction in sintering is the minimization of the total surface free energy. Actual sintering processes differ from the simplest case given above through many complicated boundary conditions. Not only outer surfaces, but also inner surfaces such as dislocations and low- and high-angle grain boundaries can serve as sources for particle fluxes. That is to say, they can serve as vacancy sinks. Furthermore, it must be noted that not only volume diffusion, but also diffusional processes along one- and two-dimensional crystal defects can lead to the elimination of pores [7,12].

2.5. Literature

General Literature:

[1] A. B. Lidiard in S. Flügge, Handbuch der Physik, Springer-Verlag, Berlin 1957, Vol. XX.
[2] W. Jost, Diffusion in Solids, Liquids, Gases, 3rd ed., Academic Press, Inc., New York 1960.
[3] K. Hauffe, Reaktionen in und an festen Stoffen, 2nd ed., Springer-Verlag, Berlin 1966.
[4] P. Kofstad, High-Temperature Oxidation of Metals, John Wiley and Sons, Inc., New York 1966.
[5] W. Seith, Diffusion in Metallen, 2nd ed., Springer-Verlag, Berlin 1955.
[6] Y. Adda and J. Philibert, La Diffusion dans les Solides, Presses Universitaires de France, Paris 1966, p. 893.
[7] R. L. Coble and J. E. Burke, Progress in Ceramic Science, Vol. III, 1963, p. 197.

Special Literature:

[8] E. W. Gorter, Philips Res. Rep. *9*, 295 (1954).
[9] A. Navrotsky and O. J. Kleppa, J. inorg. nucl. Chem. *29*, 2701 (1967).
[10] C. Wagner in Proc. 7th Meeting of the International Commission of Electrochemical Thermodynamics and Kinetics, Butterworth Scientific Publ., London 1957, p. 361.
[11] H. Rickert and C. Wagner, Ber. Bunsenges. Phys. Chemie *67*, 621 (1963).
[12] F. Thümmler and W. Thomma, Met. Rev. *115*, 69 (1967).
[13] Th. Hehenkamp, Z. Metallkunde *58*, 545 (1967).
[14] B. Ilschner, J. Chem. Phys. *28*, 1109 (1958).
[15] C. Wagner in H. Reiss and J. O. McCaldin, Progress in Solid State Chemistry, Pergamon Press, Oxford 1972, Vol. 7, p. 1.

3. Crystal defects

3.1. General remarks

The ideal crystal is an abstract concept that is used in crystallographic descriptions. The lattice of a real crystal always contains imperfections. Chemical reactions in the solid state are fundamentally dependent upon imperfections, so that an exact characterization of all possible crystalline defects is essential to an understanding of the reactivity of solids.

A suitable classification of crystalline defects can be achieved by first considering the so-called point defects and then proceeding to higher-dimensional defects. Point defects are atomic defects whose effect is limited only to their immediate surroundings. Examples are vacancies in the regular lattice, or interstitial atoms. Dislocations are classified as linear or one-dimensional defects. Grain boundaries, phase boundaries, stacking faults, and surfaces are two-dimensional defects. Finally, inclusions or precipitates in the crystal matrix can be classified as three-dimensional defects.

Following the structural description of these defects, we should make some brief remarks regarding their energies of formation. If the defect energies and their spatial distribution are known, then thermodynamics can be used to give information regarding concentrations and stabilities. In this connection, however, it is observed that only point defects are thermodynamically stable. That is, only point defects are in an equilibrium state which is uniquely determined by the specification of the requisite number of independent variables such as pressure, temperature, and composition. The concentrations and configurations of all other defects depend upon the manner in which they were introduced into the crystal. That is, in the final analysis, they depend upon the method of preparation.

3.2. Point defects

If a crystal is constructed by the stepwise addition of single particles, and if now and then a regular lattice site is left unoccupied, then we say that the crystal contains vacancies. If atoms or ions are introduced into the normally unoccupied spaces between regular lattice sites, then we speak of interstitial atoms or interstitial ions. Finally, regular lattice sites can be occupied by foreign particles. In this case we have substitutional disorder. In principle, all possible point defects have now been listed.

Defects are regions of increased energy in the crystal. In metals, alloys, and crystals which are composed of uncharged particles, the concentrations of point defects in thermodynamic equilibrium are determined by their corresponding free energies of formation. Moreover, there is no coupling between the concentrations of point defects. The situation is different in the case of ionic crystals in which the atoms are charged. In general, defects in ionic crystals possess an effective charge relative to the ideally ordered lattice. Thus, an interstitial cation is frequently positively charged, and a cation vacancy is negatively charged. In addition to the energetic considerations mentioned above, the condition of electroneutrality of the entire crystal must also be taken into account when determining the defect concentrations. As a result of this condition, the total numbers of positively and negatively charged equivalents of defects in the crystal must be equal. We can think of these defects as being grouped into neutral pairs. In the first approximation, the concentration of the defect pair with the lowest energy of formation will be so much greater than that of all other pairs that it is legitimate to speak of the disorder type of the crystal [13]. That is, the disorder type is characterized by the two oppositely charged majority defect centers. For example, in silver bromide and other silver halides we find Frenkel type disorder, in which there exist nearly equal concentrations of cation vacancies and of silver ions in the interstices [14, 28]. In alkali halides we find Schottky type disorder with nearly equal numbers of vacancies on the cationic and anionic sublattices [1, 13]. These two types of disorder are schematically shown in Fig. 3-1.

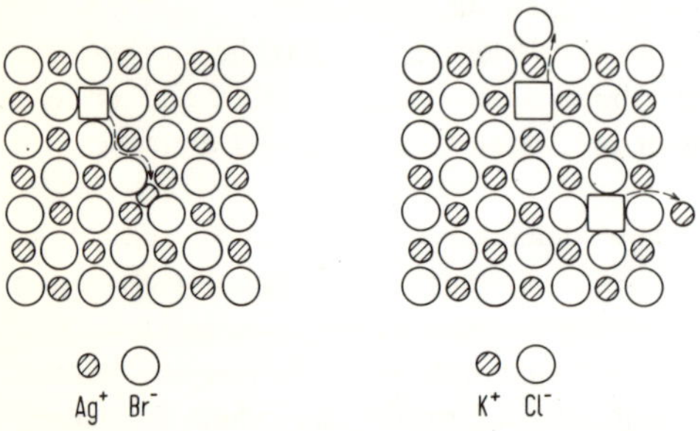

Fig. 3–1. Schematic representation of the disorder types in AgBr and in KCl. AgBr exhibits Frenkel disorder. KCl exhibits Schottky disorder. In both cases thermal disorder predominates.

A fundamentally different situation from that just described for ionic defects arises for the case of electronic defects. In metals, the density of the free electrons (conduction electrons) is so great that for many processes only the entire ensemble of electrons is manifested as a degenerate electron gas. However, for a crystal with essentially covalent or heteropolar bonding the state of affairs is completely different. In such a crystal there will be local disturbances in the periodic electrical charge distribution. These electronic disorder centers can be either excess electrons (or, simply, "electrons") or electron holes, depending upon whether there is a local charge excess or charge deficit. From the statistical thermodynamics of electronic defects which will be discussed later, it can be seen that in many cases this situation can be treated analogously to atomic point defects. Some examples of electronic defects may be given for illustration.

In silicon at 0 K, the valence band is completely filled with electrons. That is to say, each individual orbital is filled up with two bonding electrons as shown in Fig. 3-2.

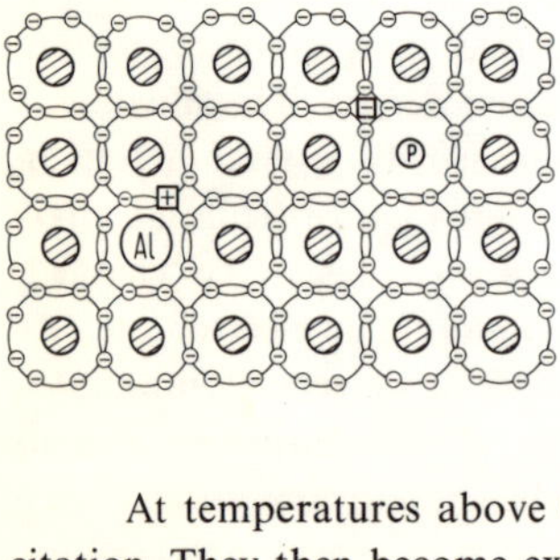

Fig. 3–2. Silicon that has been doped with phosphorus and aluminum. Electron holes ⊞ and excess electrons ⊟ exist as quasi-free particles in the vicinity of the dopant atoms Al and P which act as acceptors or donors respectively.

At temperatures above 0 K some of these electrons are set free through thermal excitation. They then become excess conduction electrons, and they belong to the conduction band of the total crystal. The remaining half-filled orbitals now become electron holes. Both electronic defects are mobile. In general, their mobilities are different and are dependent upon the crystal type. By replacing some of the silicon atoms in the crystal with atoms of 3- or 5-valent elements such as aluminum or phosphorus, the number of electron holes or excess electrons can be controlled at will, at least within the range of solubility of the dopant element [2]. This situation is also shown in Fig. 3-2. For many physico-chemical processes involving predominantly heteropolar crystals it is sufficient to think of the valence electrons, the excess electrons, and the electron holes as being relatively sharply localized on the individual ions. This is true as long as the question of their mobility is not involved [31]. For example, the nickel ions in stoichiometric nickel oxide are, of course, divalent. However, by means of a sufficiently high oxygen partial pressure, a slight excess of oxygen can be introduced, whereby, at 1 000 °C, some 10^{-4} of the divalent nickel ions become trivalent [15]. This process corresponds, in its essentials, to the introduction of electron holes in silicon by doping. What is different, however, is the stronger localization of the electronic defects in ionic crystals. Expressing it differently, one might say that an electron hole is bound for a relatively long time on a particular Ni^{2+} ion before it jumps over to one of the neighbouring nickel ions. This is a result of the relatively low mobilities of electronic defects in predominantly heteropolar crystals. Typically, these mobilities in trânsition metal oxides [3] are of the order of 1 cm^2/V sec and less. Very slow electrons in a polarizable compound will locally polarize the lattice in order to procure an energetically favourable position for themselves. The polarized region of the lattice can move along with the electron. It is then known as a polaron [16]. In comparison with the mobility of polarons, the mobilities of the electronic defects in covalent semiconductors are often several orders of magnitude higher.

As has been discussed above, atomic point defects generally carry an excess positive or negative electrical charge relative to the ideally ordered crystal. However, this charge is not merely given by the normal ionic valence in heteropolar crystals nor by the number of bonding electrons in covalent crystals. On the contrary, the atomic defects can take up electrons (give up electron holes), thereby assuming a charge state different from the usual. In this case, the defects act as acceptors. Likewise, it is possible for them to take up electron holes (give up electrons), thereby acting as donors. As an example of this, consider again an alkali halide crystal with Schottky disorder. Taking away a negatively charged chloride ion to form a chloride ion vacancy results in a positive excess charge on the vacant site relative to its undisturbed surroundings in the crystal. A conduction electron could be attracted here and held by electrostatic forces. The special configuration of an anion vacancy and a bound electron is called an F-center. (F is an abbreviation for *Farbe*, the German word for colour.) It is now known that the F-center, acting as a donor, is responsible for the absorption of visible light in alkali halides. During the absorption process, the bound electron is once again split off from the vacancy. Entirely analogous situations hold for interstitial ions or substitutional defects. Since the energy of ionization for donors and acceptors in many cases is relatively small compared to the width of the energy gap between the valence and conduction band, these defects are frequently fully ionized at temperatures at which the crystal can achieve equilibrium with respect to the atomic point defects. Multiple ionization of defects is also possible. Thus, a fully ionized Ni^{2+}-vacancy in nickel oxide bears a double negative charge relative to its surroundings. By capturing an electron hole (giving up an electron) it becomes a singly ionized vacancy. If it captures two holes (gives up to electrons) it becomes a neutral vacancy. Such different ionization steps of the cation vacancies have been observed for CoO [17, 32], although not so clearly for NiO.

For the sake of completeness it should be mentioned that donors and acceptors, as long as they are not completely ionized, can exist not only in the energetic ground state but also in exited states with respect to their electrons or electron holes. Such excited defects are called excitons. The problem of calculating the energies of the excited state is quite analogous to the same problem for free atoms or molecules.

3.2.1. Intrinsic and extrinsic defects

If the pressure P and temperature T are fixed, then the total concentrations of point defects at thermodynamic equilibrium in a compound with n components can only be dependent upon the $(n - 1)$ independent chemical potentials μ_i of the components (i.e. upon the $(n - 1)$ component activities a_i). For example, the concentration of electronic defects in silicon that has been doped with aluminum (or phosphorus) is uniquely determined by the activity of the dopant element, since this is a binary system. In a completely analogous way, the concentrations of electron holes and cation vacancies in NiO are uniquely dependent upon the oxygen partial pressure as long as overall equilibrium can be assumed.

At this point of the discussion it is worthwhile to distinguish between two different kinds of disorder. If the concentrations of the majority defect centers, which constitute the disorder type, are independent of the component activities and are only determined by P and T, then we speak of thermal disorder or intrinsic disorder (e.g. Frenkel disorder in silver bromide). However, the concentrations of minority defect centers do depend upon the component activities even in the case of a crystal with thermal disorder. This will be discussed more explicitly later. On the other hand, if the concentrations of the majority defects are dependent upon the component activities, then we speak of activity-dependent disorder or extrinsic disorder (e.g. cation vacancies and electron holes in transition metal oxides).

For both kinds of disorder the extent of the deviation from the stoichiometric composition of the compound is determined by the activity dependence of the defect concentrations. Once again, NiO will be taken as an example. Strictly, this should be written $Ni_{1-\delta}O$, where δ is the nickel deficit. That is, δ, which is dependent upon the oxygen partial pressure, is a measure of the concentration of nickel ion vacancies or of electron holes (in the form of Ni^{3+} ions). We can see, then, that it is very easy in this case to control the valence of a component by fixing an activity.

However, in a narrower sense, controlled valence has a somewhat different meaning. If lithium oxide is dissolved in nickel oxide at a fixed oxygen partial pressure, then Li^+ ions will substitute for Ni^{2+} ions. Electrically this is the same thing as introducing a charge deficit [18]. Compensation is achieved through the formation of a Ni^{3+} ion in place of a Ni^{2+} ion. For sufficiently high dopant concentrations of Li_2O, the number of Ni^{3+} ions corresponds nearly to the number of Li^+ ions introduced. This and similar situations are known as controlled valence. When looked at systematically this is nothing more than a ternary system with one atomic and one electronic majority defect center.

C. Wagner [19] was the first to point out the possibility of the controlled introduction of defects by means of dissolving a third heterovalent component in binary ionic crystals. In a binary compound, one of two things will happen, depending upon the disorder type. Either the above case of controlled valence will occur, or the excess charge introduced by the third component will be compensated by vacancies or by interstitial ions. The latter is the case, for example, when KCl with Schottky disorder or AgCl with Frenkel disorder are doped with $CdCl_2$ [1]. For a sufficient level of doping, the charge of the divalent cadmium on the cation sites will be compensated by the formation of cation vacancies which carry an effective negative charge. Simultaneously, the number of anion vacancies in KCl, or the number of interstitial cations in AgCl, will decrease because of mass action, as will be discussed later.

The considerations presented up to this point can be easily extended to higher ionic crystals and compounds with more than two or three components [4]. Again, quite generally, the energetically favourable defects constitute the disorder type. For a binary ionic crystal without electronic majority defects there are, in principle, only four disorder types. These are the previously described Schottky and Frenkel types and their corresponding anti-types: namely, cations and an equivalent number of anions in the interstices (anti-Schottky disorder), and anion vacancies with an equal number of anions in the interstices (anti-Frenkel disorder). However, for higher ionic crystals the number of possible disorder types increases considerably because of the greater number of components and sublattices. Therefore, in such crystals, it is much more difficult to uniquely determine the disorder type.

3.2.2. Associates

Point defects can interact with each other elastically through distortion of the surrounding lattice. They can also interact electrically if they are effectively charged. If the attractive forces (e.g. the coulombic interaction between oppositely charged defects) appreciably exceed the temperature energy per degree of freedom, measured in units of kT, then defect associates or defect complexes will be formed [1]. It follows that the formation of complexes is favoured by low temperatures as long as no kinetic barriers are present. Examples of complexes are: 1. double vacancies in metals [20], 2. the previously mentioned F-center which is, strictly speaking, an associate between an anion vacancy and a conduction electron [5, 21], and 3. the electrically neutral complex between a cation vacancy and a divalent alkaline earth cation in alkali halides in which cation vacancies have been introduced through doping with alkaline earth halides [22].

Crystal properties which are dependent upon the concentrations of point defects, such as the electrical conductivity in ionic crystals, are modified by the formation of complexes. Since we can conceive of a large number of possible complexes, their identification can be quite an arduous task. The most thoroughly studied defect associates are those in alkali halides. These have been examined mainly by optical methods [5, 21].

3.2.3. Energy of point defects

In order to be able to calculate the concentrations of point defects at thermodynamic equilibrium, it is necessary to know the change in free energy of the crystal which accompanies the formation of point defects, since the equilibrium is determined by the minimization of the free energy when the pressure, the temperature, and the other independent thermodynamic variables are given. A theoretical calculation of the free energy of formation of defects is still one of the most difficult problems in solid state physics and chemistry. The methods of calculation for each group of materials – metals, covalent crystals, ionic crystals – are all very different. Historically, the first energetic estimates were made for ionic crystals [23] in order to explain the different disorder types occurring in crystals with the same crystallographic structure: Frenkel disorder in silver halides and Schottky disorder in alkali halides. Then, as today, the calculations were essentially concerned with estimating the enthalpy of formation, since the entropy of formation is much less easily calculated. In order to make the fundamentals clear, and in order to have numerical values on hand for use in later thermodynamic calculations, some estimations will now be made.

In the defect thermodynamics of ionic crystals, only the free energies of formation of oppositely charged defect pairs have any significance. Also, only the energies for the formation of pairs can be experimentally determined, for instance by measuring the temperature dependence of defect concentrations. At first sight it would appear that a rough estimate of the energy of a defect pair (for example, of a Schottky pair in alkali halides) might be given by the lattice energy U_L (see eq. (1-2)). This is equal to the enthalpy change of the reaction

$$\mathrm{MeX\ (s)} = \mathrm{Me^+\ (g)_\infty} + \mathrm{X^-\ (g)_\infty}, \quad \Delta H_6\ (T = 0\ \mathrm{K}) \tag{3-1}$$

where Me, X, (s), and (g) stand for metal, metalloid, solid, and gas respectively, and where ∞ signifies the limiting case of infinite separation. ΔH_6 is experimentally accessible through the Born-Haber cycle as shown in Fig. 3-3.

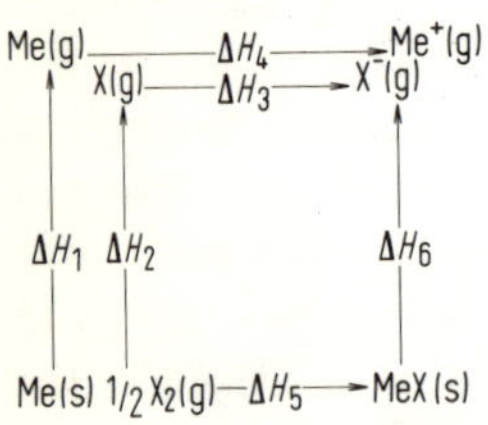

Fig. 3-3. The Born-Haber cycle to obtain the lattice energy $U_L = \Delta H_6$.

Accordingly,

$$U_L = \Delta H_6\ (T = 0\ \mathrm{K}) = \Delta H_1 + \Delta H_2 + \Delta H_3 + \Delta H_4 - \Delta H_5 \tag{3-2}$$

The assumption that the lattice energy U_L as calculated by eq. (3-2) is equal to the energy for the formation of a Schottky defect pair, as long as the lattice does not change its geometry (i.e. does not relax) during the formation of the defects, may be rationalized as follows. The removal of an Me^+ and an X^- ion from the interior of the crystal to infinity requires an energy equal to two times U_L. But by bringing these ions back from infinity to the surface of the crystal, an energy equal to U_L is recovered. Numerical values of U_L for alkali halides lie between 160 and 200 kcal/mole. This means – as will be shown in the next section – that the energies of formation of the Schottky defects calculated in this way are much too high compared to the experimentally measured values, which are about 50 kcal/mole. Thus, these values do not explain the observed defect concentrations. The way out of this situation was shown many years ago by Jost [23] who pointed out that the surroundings of a defect in ionic crystals are polarized. If covalent and van der Waals energies are neglected, the polarization energy can be estimated as follows. Suppose that the electrical charges on both members of a neighbouring anion-cation pair were neutralized by an exchange of charge in a medium of dielectric constant ε. Then the polarization energy is given as

$$u_P = \frac{e^2}{\varepsilon d}, \quad d = r_{Me^+} + r_{X^-} \tag{3-3}$$

where r_i is the ionic radius. Then, if this neutral molecule is removed to infinity in a vacuum and is ionized there, with the interionic separation maintained at d, the molecule gains back the energy

$$u_I = -e^2/d \tag{3-4}$$

Moving this molecule back from infinity to the surface of the crystal yields precisely the lattice energy U_L per mole. Thus, by putting this all together and using eq. (1-2), we obtain as a first approximation to the energy of a mole of Schottky defects:

$$U_S = U_L \left[1 - \frac{1 - \frac{1}{\varepsilon}}{\alpha_M \left(1 - \frac{1}{n}\right)} \right] \tag{3-5}$$

For sufficiently large values of ε, $1/\varepsilon$ is negligibly small compared to one. Assuming that $n = 9$ and that the Madelung constant $\alpha_M = 1.748$ (NaCl structure), we obtain $U_S \approx 0.35\, U_L$. This is of the right order of magnitude.

The treatment of polarization in the calculation of the defect energy of ionic crystals was later refined. But these refinements add nothing fundamentally new to the above treatment [24]. The preceding calculations will be familiar to physical chemists who will find them analogous to calculations of the heat of solution of ionic crystals in strongly polarizable solvents such as water. If the polarization energy is not taken into account here, then completely false results are obtained. The method of estimating the energy of a Frenkel defect pair is in principle completely analogous to the above method. It must only be noted that an additional amount of energy must be expended in order to introduce larger ions into the interstices on account of the stresses induced. Accordingly, Frenkel disorder is expected when the interstitial ions are small and when the dielectric constant ε is high so that a large polarization energy is available. This polarization energy, together with the van der Waals contribution to the bonding, can then compensate for the strain energy required at the interstitial site.

In the last few years the ways of calculating defect states, configurations, and energies in ionic crystals have become much more sophisticated, both in mathematical and computational method and in underlying physical model [33]. An immediate understanding of the nature of the point defects has thus been provided for MgO, UO_2, and CaF_2.

The discussion in this section has only been concerned with the enthalpy term. In order to determine the free energy, which is necessary for a calculation of the equilibrium defect concentration, the standard entropy change for the formation of a mole of defects may be estimated as follows. In the simplest case of the Einstein approximation for the limiting case of Dulong-Petit behaviour, the crystal with N_0 lattice atoms is considered to be a system of

3 N_0 independent oscillators that all vibrate with the same frequency ν_0. From the partition function of this ensemble of oscillators, that part of the free energy which arises from the crystal vibrations is given as [27]:

$$G_{\nu_0} = -3RT \ln \frac{h\nu_0}{kT} \tag{3-6}$$

If a defect is introduced, then the vibrational frequencies for the z neighbouring lattice atoms change from ν_0 to ν_1, so that the change in free energy accompanying the introduction of a mole of defects is of the order of

$$\Delta G_{\mathrm{v}} = -zRT \left[\ln \frac{h\nu_1}{kT} - \ln \frac{h\nu_0}{kT} \right] \tag{3-7}$$

The corresponding entropy change $-\partial \Delta G_{\mathrm{v}}/\partial T$ is therefore given by:

$$\Delta S_{\mathrm{v}} = R \ln \left(\frac{\nu_1}{\nu_0} \right)^z \tag{3-8}$$

For $z = 6$ and $\nu_1/\nu_0 = 2$ (or 1/2), numerical values of ΔS_{v} of the order of 8.4 cal/K · mole are obtained.

Little is known about estimating the point defect energy of covalent crystals or molecular crystals. In the zeroth approximation one can proceed, as for ionic crystals, via the lattice energy U_{L} which can be estimated from the enthalpy of sublimation and the zero point energy.

The point defect energy of metals consists essentially of two terms: 1. the change in the

Table 2. Vacancy formation energies and entropies for metals.

Me	$\Delta H_{\mathrm{F}}^0 \left[\frac{\text{kcal}}{\text{mole}}\right]$	$\Delta S_{\mathrm{F}}^0 \left[\frac{\text{cal}}{\text{mole K}}\right]$	Reference
Au	20.0	1.0	(1)
Pt	34.3	2.6	(2)
Ag	23	~1	(3)
Al	15.2	1.7	(4)
Cu	23.7	~0.8	(5)
Ni	32	3.0	(6)
Zn	12.2		(7)
Cd	9.2	1.0	(7)
In	13.1		(7)
Pb	11.5	1.2	(7)

[1] A. Seeger et al., Phys. Stat. Sol. *29*, 231 (1968).
[2] A. Seeger et al., Phys. Stat. Sol. *25*, 359 (1968).
[3] A. Seeger et al., Phys. Stat. Sol. *39*, 647 (1970).
[4] A. Seeger et al., Phys. Stat. Sol. *48b*, 481 (1971).
[5] A. Seeger et al., Phys. Stat. Sol. *35*, 313 (1969).
[6] A. Seeger: Vacancies and Interstitials in Metals, North-Holland Publ. Comp., Amsterdam 1969.
[7] A. Seeger, J. Phys. F: Met. Phys. *3*, 248 (1973).

Fermi energy (i.e. of the electrochemical potential η_e of the electron gas) because of the change in volume of a crystal during the formation of defects and because of the scattering of the electron waves at the defects; and 2. the change in the repulsive energy of the metal cations because of the displacement of the ions in the neighbourhood of the defects. Both terms require a knowledge of the defect volume and of the change in the configuration of the ions and electrons in the neighbourhood of the defects. The calculations can not be understood without a detailed knowledge of the methods of theoretical metal physics, and so some typical numerical values listed in Table 2 will have to suffice to give an idea of the order of magnitude of calculated and experimentally determined point defect energies in metals.

3.3. Dislocations

Dislocations are one-dimensional defects. They are largely responsible for the plastic behaviour of solids. Two of their properties are particularly important in connection with solid state reactions: 1. They can act as sites of repeatable growth within a crystal. 2. They can serve as fast diffusion paths. They also act as preferential nucleation sites for the formation of new phases.

Fig. 3–4. Edge dislocation. Burgers vector b with Burgers circuit, glide plane, and dislocation motion during plastic deformation under the action of a force F.

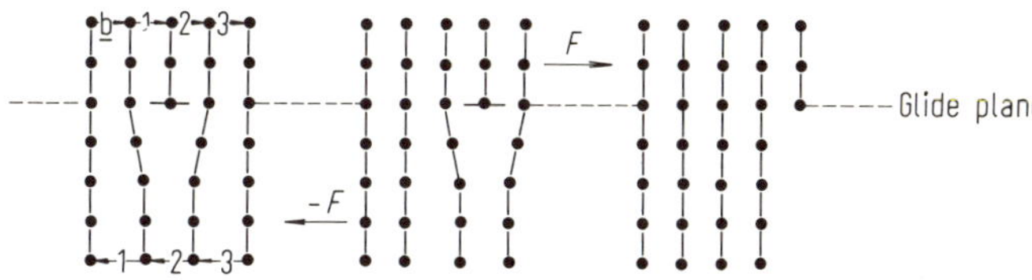

In Fig. 3–4 is shown a possible limiting case of a dislocation. This is called an edge dislocation. It can be most easily visualized as being generated by shoving a half-plane into the crystal. The edge of this plane, which is the center of the distorted region of the crystal, is called the dislocation line. Dislocations either close upon themselves or they end at inner or outer surfaces of the solid. An edge dislocation moves easily on its glide plane perpendicular to the dislocation line under the influence of a shearing force. With this can be compared the much greater force which would be required to shear an ideal crystal without dislocations. Such a glide process is shown in Fig. 3–4. A further possible mode of motion of the dislocation line perpendicular to the glide plane occurs when atoms are added to or taken away from the half-plane (climbing). The elementary steps of these additions or substractions occur chiefly at jogs in the dislocation line. The addition of particles corresponds to the removal of interstitial particles or to the formation of vacancies in the bulk crystal. The removal of particles from the half-plane corresponds to the generation of interstitial particles or to the filling up of vacancies. This so-called non-conservative motion of an edge dislocation plays a prominent role in the achievement of equilibrium among point defects, since the dislocation line serves as a site of repeatable growth in the same way as a surface does. As can be seen from Fig. 3–4, the immediate surroundings of the dislocation line are compressed on the side where the half-plane has been shoved in, and are dilated on the other side. A result of this is that point defects which possess a positive activation volume (i.e. which expand the lattice by their formation) collect in the vicinity of the dislocation line on the dilated side and reduce the stress. The opposite case holds true for the other side of the glide plane. Therefore, at equilibrium, owing to the constancy of the chemical or electrochemical potential of all constituents of the crystal, a cloud of defects is formed around the dislocation line. This anchors the dislocation to some extent (Cottrell), so that now a higher shearing stress is necessary to cause gliding during plastic deformation [25].

The second limiting case of a dislocation is shown in Fig. 3–5. This so-called screw dislocation occurs when one part of the crystal glides over another under stress so that a spiral or screw is formed in a lattice plane. The dislocation line lies in the center of the screw. Edge and screw dislocations are the limiting cases of a general dislocation line for which a part of the lattice displacement occurs perpendicular to the dislocation line as for edge dislocations, and a part occurs parallel to the dislocation line as for screw dislocations. Measured dislocation densities ϱ_D are expressed as the number of points where dislocation lines cross the crystal surface per cm^2 of surface. Values of ϱ_D are found from about 10^0 for single crystals of silicon or germanium which have been extremely carefully prepared and handled up to about 10^{12} for highly deformed metals such as copper or gold. Dislocations can be introduced into crystals in many different ways – from an accidental atomic configuration which occurs during solidification from the melt or during condensation from the vapour phase, to the bending, rolling, or cutting of a crystal. Dislocations are defects that are not in thermodynamic equilibrium.

At the end of this section it may be noted that the Burgers vector *b* is most easily visualized, as is shown in Fig. 3–4, by circulating around a dislocation line in the undisturbed lattice and counting the steps between atoms as vectors. After a complete circuit, the sum of all these vectors is called the Burgers vector *b*.

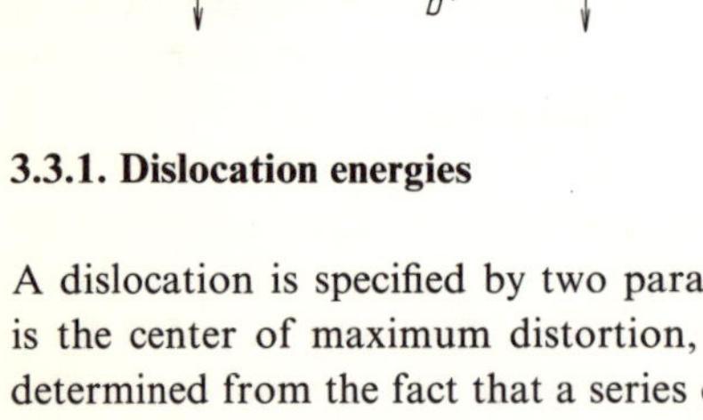

Fig. 3–5. Schematic model of a screw dislocation and its formation. The Burgers vector *b* is parallel to the dislocation line *s*.

3.3.1. Dislocation energies

A dislocation is specified by two parameters: 1. the direction of the dislocation line, which is the center of maximum distortion, and 2. the so-called Burgers vector *b*, which can be determined from the fact that a series of lattice vectors that form a closed circuit in a perfect lattice will not form a closed circuit if they surround a dislocation. The vector necessary to close the so-called Burgers circuit is the Burgers vector *b* which is expressed in units of atomic separations (see Fig. 3–4). If the Burgers vector is perpendicular to the dislocation line, then the dislocation is an edge dislocation, and if it is parallel to the dislocation line then we have a screw dislocation. The crystal lattice becomes elastically distorted in the surroundings of a dislocation [6, 7]. The field of elastic distortion of a dislocation, in contrast with that of a point defect, extends over many atomic separations and ends either at the crystal surface or in the region of influence of other dislocations. This means that for high dislocation densities the interaction energy between neighbouring dislocations becomes important. Also, in this way, point defects can interact with dislocations. If the immediate core of the dislocation of radius r_C is first cut out, then the elastic line energy E_D per cm of a screw dislocation in an isotropic medium can be easily calculated by integrating over the field of distortion of the dislocation:

$$E_D = \frac{Gb^2}{4\pi} \ln \frac{r}{r_C} \tag{3-9}$$

r is the radius of influence of a dislocation line which can be calculated in the first approximation from the dislocation density ϱ_D as $r = 1/\sqrt{\varrho_D}$. G is the shear modulus. By removing the nucleus of the dislocation, an amount of energy equal to about 10% of the line energy has been neglected. The line energy of an edge dislocation is equal to that of a screw dislocation within a factor of $(1 - \nu)^{-1}$, where ν is Poisson's ratio. Values of ν for metals frequently lie between 0.2 and 0.4. The dislocation energy is always proportional to the length of the dislocation line. Therefore, a dislocation line always strives to become shorter, and as such, a dislocation line is similar to a stretched elastic band.

By assuming a dislocation density of 10^8 per cm^2, and by assuming that r_C is of the order of 10^{-7} cm, energies E_D can be calculated from eq. (3–9) to be about 40 kcal per mole of atoms on the dislocation line. Such energies are equal to or even larger than the molar energies of single point defects. By comparison, melting energies are of the order of a few kcal/mole, and vaporization energies amount to a few tens of kcal/mole. It must be noted, however, in such comparisons, that even at this high dislocation density only about one atom in 10^7 lies on a dislocation line, which means that the contribution of all dislocations to the energy of a crystal is still quite small.

3.4. Interfaces

Interfaces play an important role in solid state reactions. During heterogeneous reactions mass transport occurs across interfaces. As sites of repeatable growth, interfaces can permit equilibrium between point defects to be attained. In sintering processes they serve as vacancy sinks and as paths of rapid transport. The interfaces which occur most frequently in crystals are outer surfaces, phase boundaries, and grain boundaries. At thermodynamic equilibrium, the electrochemical potential across an interface remains constant, but the chemical potentials of the components change because of the change in lattice structure. Therefore, electrical charges and discontinuities in electrical potential are observed at interfaces. These electrical charges can be limited to a depth of an atomic spacing as is the case for metals or, as in the case of a semiconductor, they can extend well into the interior of the material depending upon its electrical conductivity. A measure of the penetration depth of the electrical space charge layer is the so-called Debye-Hückel length [26]:

$$l = \left(\frac{\varepsilon k T}{8\pi n_i e_0^2}\right)^{1/2}$$

where ε is the relative dielectric constant, and n_i is the number per cm^3 of univalent particles with positive or negative excess charges in the lattice. If several different sorts of mobile particles i are present, n_i in the denominator must be replaced by $\sum n_i z_i^2$. The Debye-Hückel length here is taken from the theory of electrolyte solutions. A large number of interfacial phenomena are associated with the electrical surface or interface charge layer. Examples are adsorption phenomena and the growth rate of thin surface coatings during the oxidation of metals.

Systematically, a distinction is made between high- and low-angle grain boundaries. The choice of the words high- or low-angle is related only to the angle between analogous lattice planes in neighbouring crystallites. High-angle grain boundaries have a complex structure, which can extend for several atomic spacings. The mobilities of particles in grain boundaries, and particularly of impurities which collect in grain boundaries, are in most cases much larger than in the bulk lattice. Thus, polycrystallinity can have a definite influence on the reactivity of solids. It can be appreciated that high-angle grain boundary effects are very difficult to characterize quantitatively on account of the infinite variety of this type of interface. This is not the case, however, for low-angle grain boundaries in single-phase material. Two cases can be distinguished here. Firstly, if two parts of a crystal are rotated relative to each other through only a small angle, and if the axis of rotation lies in the interface, then the low-angle grain boundary is a collection of edge dislocations as shown in Fig. 3-6. The separation, a, of the half-planes which have been shoved in determines the angle of rotation $\theta \propto 1/a$.

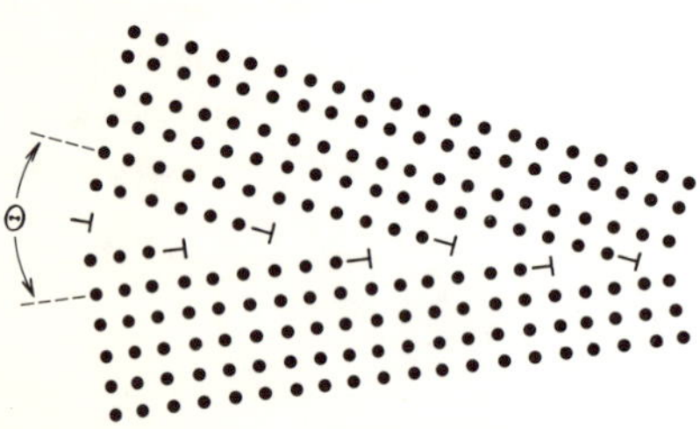

Fig. 3–6. Model of a low-angle grain boundary comprised of edge dislocations. $\theta \simeq b/a$ for the case where $b \ll a$. b is the Burgers vector, and a is the separation of the individual lattice planes which have been shoved in.

Secondly, a low-angle grain boundary can be comprised of a collection of screw dislocations which cause two parts of a crystal to undergo a small rotation relative to one another with the axis of rotation perpendicular to the low-angle grain boundary. Since the energy and interaction of dislocations are quantitatively well-understood, it is possible to treat low-angle grain boundaries theoretically.

If two parts of a single-phase crystal are identical and are coherently joined to one another with no distortion, then we have a twinning plane. This is only possible for certain lattice planes and for certain definite orientations of the two parts of the crystal relative to one another. The grain boundary in this case is a mirror plane.

Stacking faults are yet another type of two-dimensional crystal defects. These are irregularities in the stacking of certain lattice planes. As an example, let us consider the sequence of the (111)-planes of a close-packed face-centred-cubic crystal. As one can easily verify, the individual (111)-planes have an hexagonal structure, and each successive plane is displaced relative to the previous one by one spherical radius along both axial directions of the hexagonal planes, so that after a total of three planes have been stacked, the fourth plane lies directly over the first. If the three possible positions of these planes are designated A, B, and C, then the normal sequence is . . . ABCABC. . . , and a stacking fault occurs when the stacking is altered, as for instance ...ABCABABC.... The four successive planes ABAB comprise the basic unit of an hexagonal-closest-packing. Therefore, stacking faults can easily arise if the difference in energy between the hexagonal and the cubic-face-centred lattices is small. Stacking faults can occur through migration of dislocations in glide planes.

In multiphase polycrystalline material, principally the same interfaces are found between the various phases as are found between crystallites in single-phase solids. The formation of interfaces depends in each case on the degree of similarity in crystal structure and interatomic distance. This will determine whether the phase boundaries are coherent, partially coherent, or incoherent (see also Fig. 7–9).

The occurrence of all those interfaces which have been mentioned depends upon the conditions of preparation of the solid – as, for example, upon the method of solidification or of deposition from the vapour phase, or upon rolling, drawing, bending, etc. – and upon the subsequent annealing process by means of which transformation, recrystallization, or relaxation (i.e. the formation of low-angle grain boundaries at the expense of randomly distributed dislocations) can proceed [7].

3.4.1. Interfacial energies

In order to bring atoms reversibly and isothermally from the undistorted interior of a crystal to an interface – particularly to the surface – an expenditure of free energy is required. This is best illustrated by the following example. An atom in a close-packed solid metal has 12 nearest neighbours. After melting, it has on the average about 11 nearest neighbours, and in the ideal gas phase it has no interacting neighbours. Furthermore, as can be easily seen by counting the bonds, 9 bonds, or 18 half bonds, to nearest neighbours are broken during sublimation, and a surface atom possesses 3 half bonds which are free to the gaseous atmosphere. Therefore, the surface energy E_0 corresponds to about 1/6 of the heat of sublimation. It also follows that the interfacial solid-liquid energy should be about 1/12 of this amount. Today there are sophisticated methods which can be used to calculate the energy of a free surface with the help of a potential vs. interatomic distance curve (see Fig. 1-2) [8]. However, our purpose here is only to provide an understanding of the phenomena and to show the order of magnitude of the effects, as is done is Table 3. Therefore, these few brief comments will have to suffice.

Table 3. Surface energies E_0 for various solids from J. J. Gilman, J. Appl. Phys. *31*, 2208 (1960).

	E_0 (exp) (ergs/cm^2)	E_0 (theor) (ergs/cm^2)
NaCl	300	310
LiF	340	370
MgO	1200	1300
CaF_2	450	540
BaF_2	280	350
$CaCO_3$	230	380
Si	1240	890
Zn	105	185
Fe (3% Si)	1360	1400

High-angle grain boundaries can be most closely compared with liquid films. Accordingly, their energy should be of the order of magnitude of that of liquid-crystal interfaces (i.e. about 10^2 ergs/cm^2). Values of this order are actually observed. It should be mentioned here that computational methods have been recently developed which allow the strict calculation of grain boundary energies of metals if the grains are arbitrarily oriented with regard to each other. Finally it should be noted that the estimates are concerned with giving values of the energies, and not of the free energies. For large enough mobilities of the particles in the surface, the latter can be measured as surface tensions.

The interfacial energy of low-angle grain boundaries which consist of arrays of dislocations (see Fig. 3-6) can be calculated from the elastic energy of interaction between the dislocations which form the grain boundary. In eq. (3-9) the energy of an individual dislocation E_D was given. The number of dislocations per unit area between two crystallites whose lattice planes are at angle θ as in Fig. 3-6 is given by $1/a = \theta/b$. Furthermore, the radius of influence r of an edge dislocation in the low-angle grain boundary must be a fraction of a, and so all together the energy of the low-angle grain boundary per unit area is given as:

$$E_{\text{low}} = \frac{G b \theta}{4 \pi (1 - \nu)} (\ln \beta - \ln \theta) \tag{3-10}$$

where β is a constant which can be calculated from the considerations given above, and ν again is Poisson's ratio. For angles up to $\theta \simeq 6°$ the theoretical calculations are substantiated by experiments. Energies of low-angle grain boundaries are of the order of a power of ten lower than the energies of high-angle grain boundaries. However, as shown in eq. (3–10), the values are strongly dependent upon the difference in orientation θ between the mosaic blocks [29].

If a dislocation line with Burgers vector b splits into two lines with Burgers vectors b' and b'', and if the latter are not lattice vectors, then a planar lattice defect is formed between the two new dislocation lines. Such lattice defects are identical to the stacking faults described in section 3.3. They will be stable if $b^2 > b'^2 + b''^2$, because the energy E_D of an individual dislocation is proportional to the square of the Burgers vector b as shown in eq. (3-9). The energies of stacking faults are of the order of some 10 ergs/cm^2. The separation of the partial dislocations which generated the stacking fault amounts to some 10 atomic separations.

3.5. Three-dimensional crystal defects

Pores and macroscopic inclusions are three-dimensional crystal defects. From the standpoint of the reactivity of solids, pores can be very important. Consider, for instance, the formation of porous scales during oxidation (tarnishing) [11]. (For example, the decarburization of iron cannot occur if a non-porous oxide scale without grain boundaries is formed on its surface.) Or consider the direct reduction of ore [10] in which the reduction rate is greatly dependent upon the formation of porous metal surface layers. In many so-called solid state reactions, gaseous products are formed as well as solid reaction products as, for example, during the reaction of TiO_2 with $BaCO_3$ to produce $BaTiO_3$ with the formation of $CO_2^{(g)}$. In such cases, just as in the case of ore reduction, the formation of a porous product surface layer is of decided importance for the progress of the reaction.

Pores may be classified according to morphology:

1. Closed isolated pores which, during transport processes through the crystal, can act essentially either as infinitely large resistances or as short circuits, depending upon the vapour pressure and rate of vaporization of the component being transported.

2. Open pores, which are connected with each other and also with the surface of the crystal. The form and distribution of the pores are most important for transport phenomena such as diffusion, Knudsen flow, and surface diffusion. Because of the technological importance of this field, it possesses an extensive literature, especially on procedures for determining porosity. We can only make mention of this literature here [12].

Three-dimensional inclusions occur in solid phases in several forms. First of all there are impurities which separate out either as gaseous or as solid phases during the preparation of solids from the melt. Solid inclusions can have either a coherent or an incoherent phase boundary with the matrix, depending upon how widely the phases differ from one another in structure and lattice constants. Inclusions are undesirable if they disturb the homogeneity of the structure, as in the case of steel. On the other hand, their resistance to plastic deformation can be used to technological advantage in dispersion hardening and in the so-called age-hardening alloys [30]. In the latter case, two- or three-dimensional inclusions precipitate in the matrix in a controlled manner when the solubility limit of an added element is surpassed. Because of their stress field and their different structure, the inclusions interact with the dislocations and hinder their migration [9], thus causing an increase in hardness.

3.6. Literature

General Literature:

[1] A. B. Lidiard in S. Flügge, Handbuch der Physik, Springer-Verlag, Berlin 1957, Vol. XX, p. 258.
[2] E. Spenke, Elektronische Halbleiter, 2nd ed., Springer-Verlag, Berlin 1965.
[3] H. Eggert in F. Sauter, Festkörperprobleme I, Halbleiterprobleme VII, Friedr. Vieweg and Son, Braunschweig 1962, p. 274.
[4] H. Schmalzried in H. Reiss, Progress in Solid State Chemistry, Pergamon Press, Oxford 1965, Vol. 2, p. 265.
[5] F. A. Kröger, The Chemistry of Imperfect Crystals, North-Holland Publ. Comp., Amsterdam 1964, p. 522.
[6] R. W. Cahn, Physical Metallurgy, North-Holland Publ. Comp., Amsterdam 1965, p. 621.
[7] P. G. Shewmon, Transformations in Metals, McGraw-Hill Book Comp., Inc., New York 1969.
[8] A. W. Adamson, Physical Chemistry of Surfaces, 2nd ed., John Wiley and Sons, New York 1967, p. 272.
[9] A. Kelly and R. B. Nicholson, Precipitation Hardening, Pergamon Press, London 1963.
[10] L. von Bogdandy and H.-J. Engell, Die Reduktion der Eisenerze, Springer-Verlag, Berlin 1967.
[11] H. Pfeiffer and H. Thomas, Zunderfeste Legierungen, 2nd ed., Springer-Verlag, Berlin 1963.
[12] H. Salmang and H. Scholze, Die physikalischen und chemischen Grundlagen der Keramik, 5th ed., Springer-Verlag, Berlin 1968, p. 103.

Special Literature:

[13] W. Schottky, Z. phys. Chem. (B) *29*, 335 (1935).
[14] J. Frenkel, Z. Physik *35*, 652 (1926).
[15] Y. D. Tretjakov and R. A. Rapp, Trans. AIME *245*, 1235 (1969).
[16] H. Fröhlich, Adv. Physics *3*, 325 (1954).

[17] B. Fisher and D. S. Tannhauser, J. Electrochem. Soc. *111*, 1194 (1964).
[18] E. J. W. Verwey, P. W. Haaijman, F. C. Romeijn and G. W. van Oosterhout, Philips Res. Rep. 5, 173 (1950).
[19] E. Koch and C. Wagner, Z. phys. Chem. (B) *38*, 295 (1937).
[20] R. W. Cahn, Physical Metallurgy, North-Holland Publ. Comp., Amsterdam 1965, p. 688.
[21] R. W. Pohl, Proc. Phys. Soc. (London) *49*, 1 (1937).
[22] F. Bassani and F. G. Fumi, Nuovo Cim. *11*, 274 (1954).
[23] W. Jost, J. chem. Phys. *1*, 466 (1933).
[24] E. S. Rittner, R. A. Hutner and F. K. du Pré, J. chem. Phys. *17*, 198 (1949).
[25] A. H. Cottrell and M. A. Jaswon, Proc. Roy. Soc. (London) *A 199*, 104 (1949), Phil. Mag. *43*, 645 (1952).
[26] P. Debye and E. Hückel, Physik. Z. *24*, 185 (1923).
[27] J. D. Fast, Entropie, Philips Techn. Bibliothek, Eindhoven 1960, p. 152.
[28] K. Weiss, Z. phys. Chem. NF *67*, 86 (1969).
[29] K. Lücke, Z. Metalikunde *44*, 370 (1953).
[30] Collected articles: Dispersionshärtung, Freiberger Forschungshefte B 142, VEB Deutscher Verlag für Grundstoffindustrie, Leipzig 1969.
[31] A. J. Bosman and H. J. van Daal, Adv. Physics *19*, 1 (1970).
[32] G. Schwier, R. Dieckmann, and H. Schmalzried, Ber. Bunsenges. physik. Chemie *77*, 402 (1973).
[33] A. B. Lidiard and M. J. Norgett in F. Hermann, Computational Solid State Physics, Plenum Press, New York 1972, p. 385.

4. Thermodynamics of point defects

4.1. General remarks

The reactivity of solids is brought about almost entirely as a result of the disorder in crystals. The most important lattice defects in connection with chemical reactions are point defects. In order that a chemical reaction may take place in a finite time, it must be carried out above a certain minimum temperature, where the defects which give rise to transport have a sufficiently large mobility. Therefore, in most cases it can be assumed that local defect equilibrium is attained during a reaction, as long as there are sufficient sources and sinks for point defects.

Despite their relatively high enthalpy of formation, point defects are the only defects which exist in thermodynamic equilibrium in any appreciable concentrations. As shown in section 3.3.1, about the same amount of energy is required per atom on a dislocation line as is required for a point defect. However, one dislocation line can contain about 10^7 such atoms, and the thermodynamic probability of defects being lined up along a dislocation line, as opposed to being freely distributed in the crystal, is very small. Thus, it can easily be appreciated that the probability of the occurrence of equilibrium dislocations is negligible.

The equilibrium concentrations of point defects are fixed by pressure, temperature, and by the composition of the crystal. Since in most cases their concentrations are very small, the laws of dilute solutions may be used to calculate the dependence of the concentrations of defects upon the independent thermodynamic variables. Deviations from ideally dilute conditions, especially because of electrostatic interaction between electrically charged point defects, will be considered in greater detail later in connection with phases which have a wide range of homogeneity. If the problem of calculating the concentrations of the point defects can be solved by using classical thermodynamics and some equations of statistical thermodynamics, then, in principle, the transport coefficients for diffusion-controlled solid state reactions can be determined as functions of the local composition of the reaction product crystal. It then becomes possible to treat quantitatively many reaction problems in the solid state. Even for reactions which are not diffusion-controlled, it will be shown that the rate constants of the reaction are determined to a large extent by point defects. For all these reasons, then, a concise summary of defect thermodynamics will be given in this chapter.

4.1.1. Temperature dependence of defect concentrations

According to Nernst's postulate, completely ordered solid phases (i.e. ideal crystals) occur when equilibrium is attained at 0 K. However, at finite temperatures the concentrations of point defects are fixed through the condition of minimization of the free energy. For given values of pressure, temperature, and component activities, these concentrations are dependent upon the magnitude of the free energy of formation of the defects. As an example, let us briefly consider the temperature dependence of the vacancy concentration in a crystal composed only of A atoms. If N_i are the mole fractions, then the free enthalpy G_m per mole of A lattice sites is given by $\sum_i N_i \mu_i$ where i stands for atoms A and for vacancies V, and where μ_i is the partial molar free energy of i. It is assumed that the vacancy concentration is very small, so that the interaction between the individual vacancies can be neglected. Then G_m is explicitly given as ($N_A = 1 - N_V$)

$$G_m = (1 - N_V)\,\bar{H}_A + N_V \bar{H}_V - T\,[(1 - N_V)\,\bar{S}_A + N_V \bar{S}_V] - TS^C \tag{4-1}$$

where $\bar{H}_i$ and $\bar{S}_i$ are the partial molar enthalpy and vibrational entropy, and S^C is the configurational entropy. Because of the assumed absence of interaction between the defects, it may be further assumed that they are randomly distributed, and so S^C is equal to the ideal entropy of mixing

$$S^C = -R\,[(1 - N_V)\ln(1 - N_V) + N_V \ln N_V] \tag{4-2}$$

If the minimum value of G_m at given P and T is now sought by setting the derivative $(\partial G_m/\partial N_V)$ equal to zero, it follows from eqs. (4-1) and (4-2) that

$$N_V = \exp\frac{\Delta S_V}{R} \cdot \exp -\frac{\Delta H_V}{RT} \simeq \text{const} \cdot \exp -\frac{\Delta H_V}{RT} \tag{4-3}$$

where $\Delta H_V = \bar{H}_V - \bar{H}_A$ and $\Delta S_V = \bar{S}_V - \bar{S}_A$. The operator Δ thus signifies the formation reaction of a mole of vacancies which are formed by removing A atoms. Eq. (4-3) states that, under the given assumptions, the defect concentration varies as the exponential of the reciprocal of the absolute temperature. Eq. (4-3) essentially describes the defect concentration dependence in metals.

4.1.2. Symbolism of defects

A glance at binary or ternary phase diagrams will show that solution phases with regions of existence of every possible extent of composition can be found [1, 2]. In this context there is no special position held by so-called stoichiometric compounds. They can be understood as limiting cases of compounds with very narrow regions of homogeneity. In the following discussion, especially in case of compounds with narrow ranges of homogeneity, it will be preferable to take P, T, and the component activities a_i as independent variables rather than P, T, and the mole fractions N_i. The point defect concentrations, which determine the deviations from the stoichiometric composition, are then dependent upon the component activities. If the defects are considered to be quasi-particles as in section 4.1.1, then the dependence of the concentrations of these particles upon the component activities can be calculated by means of solution thermodynamics. The fact that defect thermodynamics has up until today essentially only been applied to compounds with narrow regions of homogeneity is due to the fact that the limiting laws of ideal dilute solutions have almost always been used for defects in solids [3].

All statements which have been made up to now also apply for ionic crystals. At first sight it would seem that we do not have the prerequisites necessary for a definition of the chemical potentials of the defects in ionic crystals relative to convenient standard states, since firstly the electroneutrality of the crystal must be preserved, and secondly the proper relationships between the number of crystallographic different lattice sites must be preserved. However, as shown in section 3.2, in ionic crystals only those defects occur which give rise to electrically neutral pairs. This automatically maintains the correct relationship between the lattice sites, and so we do in fact have enough variables to fix the chemical potentials of the defects [4]. In order now to formulate the equilibria between the independent components of the crystal and the defects, a symbolism for particles must be developed which goes beyond the usual symbolism for species in normal chemical reactions.

The symbols which we introduce for the structural elements of a crystal, including the defects, should express the following three characteristics of a structural element: 1. what the structural element is (atom, vacancy, etc.), 2. the position in the crystal lattice, and 3. the electrical charge of the structural element. There are two possible ways to designate the electrical charge: a) it can be given relative to the ideal crystal taken as standard state, or b) the absolute charge can be given. Structural elements do not have chemical potentials which can be independently predetermined. Independent chemical potentials can only be given for combinations of structural elements. These have been called building units (Bauelemente) [5]. In the following discussion, the Kröger symbolism will be used [4]. The symbol for a structural element is given as S_P^C where: S = species (including vacancies for which we write V), C = = charge, and P = crystallographic position. For example, some structural elements of silver bromide may be listed: Ag_{Ag}^{x}, $Ag_i^{\cdot}$, V_{Ag}', V_i^{x}. The superscript $^{\cdot}$ stands for a positive excess charge, $'$ for a negative excess charge, and ^{x} indicates neutrality relative to the ideal crystal. The subscripts Ag and i mean regular silver ion sites and interstitial sites.

In formulating reactions in which structural elements take part, one must observe three balances: 1. the material balance, 2. the charge balance, and 3. the site balance. It is understood that the ratios to one another of crystallographically possible sites are not permitted to change for the formulated reaction.

4.2. Equilibrium conditions

After these introductory remarks, let us calculate the equilibrium concentrations of structural elements in a binary crystal MeX when the independent variables P, T, and the chemical potential μ_i of a component (Me or X) are given. From the Gibbs phase rule it is obvious that with P, T, and μ_{Me} (or μ_X) fixed, all independent variables are determined. That is, the concentrations of all defects can be calculated as functions of these independent variables.

In an isothermal and isobaric system in which the total pressure P is fixed by an inert gas, the solid crystal exists in equilibrium with its gaseous components. The differential of the free energy of this system is:

$$dG = \sum_{s} \mu_i dn_i + \sum_{g} \mu_j dn_j \tag{4-4}$$

where s indicates summation over all structural elements of the solid phase, and g indicates summation over all gaseous constituents. Let us consider first an arbitrary reaction between a component of the crystal MeX in the gas phase and certain structural elements of the crystal, as for example:

$$1/2\, X_2\,(g) + 2\,Me_{Me}^{x} = MeX\,(g) + Me_{Me}^{\cdot} + V_{Me}' \tag{4-5}$$

At equilibrium, the derivative of G with respect to the advancement of the reaction $\mathrm{d}\xi = \mathrm{d}n_i/v_i$ is equal to zero (where v_i are the stoichiometric factors in the reaction equation). Thus, it follows directly from eq. (4-4) that

$$\sum_{s} v_i \mu_i + \sum_{g} v_j \mu_j = 0 \tag{4-6}$$

By substituting this relationship into the reaction equation (4-5), and noting the definition of the chemical potential $\mu_i = \mu_i^0 + RT \ln a_i$, it may be seen from eq. (4-6) that

$$\frac{a_{V'_{Me}} \cdot a_{Me^{\bullet}_{Me}}}{p_{X_2(g)}^{1/2}} = \exp - \frac{\Delta G_5^0}{RT} = K_5 \tag{4-7}$$

where ΔG_5^0 is the standard free energy change of reaction (4-5), and K_5 is the equilibrium constant of this reaction. In proceeding from eq. (4-6) to eq. (4-7), two points should be noted. 1. Because of the coexistence of the gas phase with the pure material MeX in its standard state, the activity a_{MeX} (g) is unity. 2. The concentration of Me^{x}_{Me} is always very large in comparison to the defect concentrations, and it changes only imperceptibly following a change of the independent variables. Therefore, Me^{x}_{Me} can always be taken as being in the standard state, so that it does not appear explicitly in eq. (4-7). Then, by applying the limiting laws of ideal dilute solutions to eq. (4-7) it follows that

$$N_{V'_{Me}} \cdot N_{Me^{\bullet}_{Me}} = K_5 \cdot p_{X_2(g)}^{1/2} \tag{4-8}$$

Furthermore, as can be seen from reaction (4-5), the stipulations of material, site, and charge balance are fulfilled. In all the following discussions it will be simplest to give the concentrations of defects relative to a lattice molecule (i.e. relative to a formula unit of crystalline MeX). This will be indicated by placing the symbol for the defect in round brackets so that eq. (4-8) may be written

$$(V'_{Me}) \cdot (Me^{\bullet}_{Me}) = K_5 \cdot p_{X_2(g)}^{1/2} \tag{4-8a}$$

From eqs. (4-5) and (4-8) it can be seen that formal mass action equations may be written for the concentrations of defects in equilibrium when P, T, and the activity or partial pressure of a component are given. As has already been pointed out, the formulation of the reaction equation connecting the defect concentrations and the component activities is to some degree arbitrary, as long as the previously mentioned rules are properly obeyed. This results from the existence of internal defect equilibria which are always maintained if the crystal is in thermodynamic equilibrium. In the present case, for instance, we can write several other reaction equations. For example, we could write:

$$V_i^{x} + Me^{x}_{Me} = Me^{\bullet}_{i} + V'_{Me} \tag{4-9}$$

where the subscript i indicates interstitial sites. From this equation we can derive the condition for Frenkel disorder equilibrium in the same manner as above:

$$(Me^{\bullet}_{i}) \cdot (V'_{Me}) = K_9 \tag{4-10}$$

In general it can be shown that exactly enough internal defect equilibria exist to permit us to express all defect concentrations as functions of the independent variables, as long as the material, site, and charge balances are observed. This explains the free choice which we have in formulating eq. (4-5) (i.e. in formulating the external equilibrium conditions).

4.2.1. The disorder in silver bromide

In practice, the explicit calculation of the concentrations of all defects as functions of the independent thermodynamic variables (which is the basic problem in defect thermodynamics) in this general form is not an easy matter. However, it is also mostly unnecessary [4]. The reason for this is that by introducing disorder types (i.e. majority and minority defects) we can neglect the concentrations of the minority defects in the balance equations and in the conditions of electroneutrality. The various relationships then become immediately clear.

In the following example concerning the equilibrium disorder in AgBr we shall illustrate the principles which we have been discussing, and we shall illustrate the methods by which such problems may be treated. The methods will also apply for higher ionic crystals where only the number of components, and therefore the number of external equilibrium conditions, is increased.

The disorder in AgBr will be discussed as a function of the component activities at given P and T. Because of the reaction equation $Ag + 1/2\ Br_2 = AgBr, (\Delta G^0_{AgBr})$, the activities of Ag and Br_2 in AgBr are coupled by the formula

$$a_{Ag} p_{Br_2}^{1/2} = \exp \frac{\Delta G^0_{AgBr}}{RT} \tag{4-11}$$

Therefore, it makes no fundamental difference which of these two activities is chosen as the independent variable. In practice, p_{Br_2} is easily fixed. Therefore we shall choose this variable for the formulation of the reaction equations and the corresponding equilibria. This can be generalized as follows. In practice, the partial pressure of the electronegative component is frequently an easily controllable and determinable variable. Furthermore, silver bromide is always found in the standard state, since deviations from the stoichiometric composition AgBr are not chemically measurable for all values of the independent variables. Possible defects are: $Ag_i^{\bullet}$, $Ag_{Br}^{\bullet\bullet}$, V'_{Ag}, Br'_i, Br''_{Ag}, $V_{Br}^{\bullet}$, and finally, the electronic defects e' and $h^{\bullet}$ (i.e. excess electrons and electron holes which can be identified in the particle terminology with Ag'_{Ag} and $Br_{Br}^{\bullet}$). For the purpose of listing the possible disorder types (i.e. combinations of majority

Table 4. Scheme for ascertaining the possible disorder types in the ionic crystal AgBr. Actual listings are for all possible types with purely thermal disorder.

	V'_{Ag}	Br'_i	Ag'_{Ag}	Br''_{Br}
$Ag_i^{\bullet}$	$(V'_{Ag}) = (Ag_i^{\bullet})$	$(Br'_i) = (Ag_i^{\bullet})$		
$V_{Br}^{\bullet}$	$(V'_{Ag}) = (V_{Br}^{\bullet})$	$(Br'_i) = (V_{Br}^{\bullet})$		
$Br_{Br}^{\bullet}$				
$Ag_{Br}^{\bullet\bullet}$				

disorder centers in the sense of section 3.2), Table 4 is given. All structural elements with negative excess charge are given a row in this table, and all those with positive excess charge are given a column. All possible disorder types can be read off at the intersection of a row and a column. The four disorder types which are actually listed in Table 4 are those which are possible for the case of purely thermal disorder for exactly stoichiometric AgBr. All other boxes designate disorder types for which AgBr has either an excess or a deficit of Ag. These have no special names.

As external equilibrium condition (i.e. as the relationship between the chosen independent variable p_{Br_2} and the dependent defect concentrations at fixed P and T), the following reaction equation between components and structural elements may be written:

$$1/2\ Br_2\ (g) + Ag^x_{Ag} = AgBr\ (g) + V'_{Ag} + h^{\cdot} \tag{4-12}$$

The corresponding equilibrium condition is:

$$(h^{\cdot}) \cdot (V'_{Ag}) = K_{12} \cdot p^{1/2}_{Br_2} \tag{4-13}$$

The concentrations of $Ag^{\cdot\cdot}_{Br}$ and Br''_{Ag} can be neglected since these defects are very unlikely on account of their high electrostatic energies. Thus, there are eight concentrations of structure elements which must be calculated as functions of the component activities at fixed P and T. These are: (Ag^x_{Ag}), (Br^x_{Br}), $(Ag^{\cdot}_i)$, (Br'_i), (V'_{Ag}), $(V^{\cdot}_{Br})$, (e'), and $(h^{\cdot})$. The necessary equations are given by the various material, site, and charge balances and by the internal equilibrium conditions. The site balance, again relative to the lattice molecule, is as follows:

$$(Ag^x_{Ag}) + (V'_{Ag}) + (e') = 1; \quad e' \equiv Ag'_{Ag} \tag{4-14}$$

$$(Br^x_{Br}) + (V^{\cdot}_{Br}) + (h^{\cdot}) = 1; \quad h^{\cdot} \equiv Br^{\cdot}_{Br} \tag{4-15}$$

The material balance for $Ag_{1+\delta}Br$, where δ is the deviation from the exact stoichiometric composition, yields the equations:

$$(Ag^x_{Ag}) + (Ag^{\cdot}_i) + (e') = 1 + \delta \tag{4-16}$$

$$(Br^x_{Br}) + (Br'_i) + (h^{\cdot}) = 1 \tag{4-17}$$

The electroneutrality condition is:

$$+1 \cdot [(Ag^{\cdot}_i) + (V^{\cdot}_{Br}) + (h^{\cdot})] - 1 \cdot [(Br'_i) + (V'_{Ag}) + (e')] = 0 \tag{4-18}$$

Finally, for the internal equilibria between the various structural elements we have the following reaction equations and the corresponding equilibrium conditions:

$$V^x_i + Ag^x_{Ag} = Ag^{\cdot}_i + V'_{Ag} \qquad (Ag^{\cdot}_i) \cdot (V'_{Ag}) = K_{19} \tag{4-19}$$

$$V^x_i + Br^x_{Br} = Br'_i + V^{\cdot}_{Br} \qquad (Br'_i) \cdot (V^{\cdot}_{Br}) = K_{20} \tag{4-20}$$

$$e' + h^{\cdot} = 0 \qquad (e') \cdot (h^{\cdot}) = K_{21} \tag{4-21}$$

Eqs. (4-14) to (4-21) are the eight equations which can be used to calculate the concentrations of the eight structural elements mentioned above as functions of the component activity p_{Br_2}.

From electrical conductivity measurements, from transference measurements, and by combining lattice constant and density measurements, it has been shown that Frenkel disorder predominates in pure undoped AgBr [6]. Estimates of disorder energies as in section 3.2.3 can be made. They confirm the experiments. For purposes of solving the system of equations (4-13) to (4-21), however, this means that only (V'_{Ag}) and $(Ag_i^{\cdot})$ need to be considered in the balance equations. These are the concentrations of the majority defect centers which constitute the disorder type. All other concentrations (of the minority defect centers) can be neglected. Since δ is not chemically measurable (i.e. $\delta \ll 1$), it follows from eqs. (4-14) and (4-16) that $(V'_{Ag}) \simeq (Ag_i^{\cdot})$. The relative partial pressure of bromine $p_{Br_2}/p_{Br_2}(\delta = 0)$ can be calculated from eqs. (4-12) to (4-21) as:

$$\left(\frac{p_{Br_2}}{p_{Br_2(\delta=0)}}\right)^{1/2} = \left[\frac{|\delta|}{2\alpha} + \sqrt{1 + \left(\frac{\delta}{2\alpha}\right)^2}\right] \cdot \left[\frac{|\delta|}{2\beta} + \sqrt{1 + \left(\frac{\delta}{2\beta}\right)^2}\right] \tag{4-22}$$

where $\alpha = K_{19}^{1/2}$ and $\beta = K_{21}^{1/2}$. Since $\alpha \gg \beta$ for a pure material with Frenkel disorder, and since $\delta \ll \alpha$, the first bracketed expression in eq. (4-22) can be set equal to unity. The remaining expression for the limiting case $|\delta| \gg \beta$ is easily understood. One obtains in logarithmic form:

$$\frac{1}{2}\log\frac{p_{Br_2}}{p_{Br_2(\delta=0)}} = \log\delta - \log\beta \text{ (or } \log\beta - \log|\delta|) \tag{4-22a}$$

Eq. (4-22) is graphically presented in Fig. 4-1.

For $\delta = 0$ there is a point of inflection. Moreover, the curve is similar to a titration curve, and the slope of the curve at the inflection point is given by the equilibrium constant

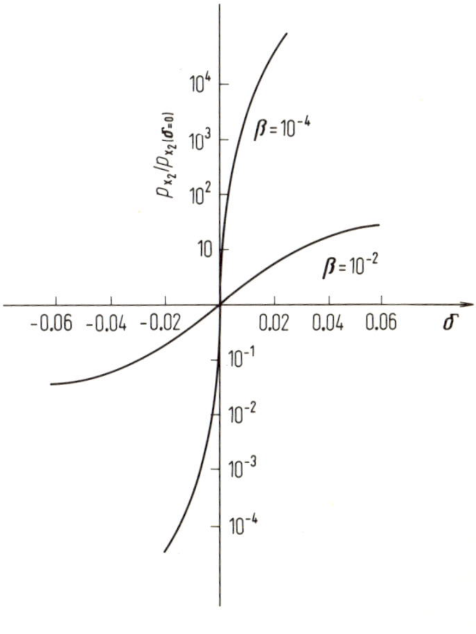

Fig. 4-1. Schematic graph of the relative partial pressure of the electronegative element of a binary compound as a function of the deviation δ from the stoichiometric composition (see also [19]).

$\beta = K_{21}^{1/2}$. The larger this quantity is, the flatter is the "titration curve". The analogy to the titration process is much more than just a formalism [4]. All further relationships can now be easily derived. Only the most important equations will be mentioned. These are:

$$(V'_{Ag}) = (Ag_i^{\bullet}) = K_{19}^{1/2} \tag{4-23}$$

$$(h^{\bullet}) = K_{12} \cdot K_{19}^{-1/2} \cdot p_{Br_2}^{1/2} \tag{4-24a}$$

$$(e') = K_{19}^{1/2} \cdot K_{12}^{-1} \cdot K_{21} \cdot p_{Br_2}^{-1/2} \tag{4-24b}$$

Therefore, the concentrations of electrons and electron holes as minority defects should depend in a predictable way upon the partial pressure of bromine or, from eq. (4-11), upon the silver activity in AgBr. This relationship is most easily tested experimentally by means of electrical conductivity measurements.

Despite the low concentration of electronic charge carriers, the electronic conductivity is measurable because the mobility of the electronic defects is orders of magnitude higher than that of the ionic defect centers. The partial conductivity values relative to each other are shown schematically in Fig. 4-2.

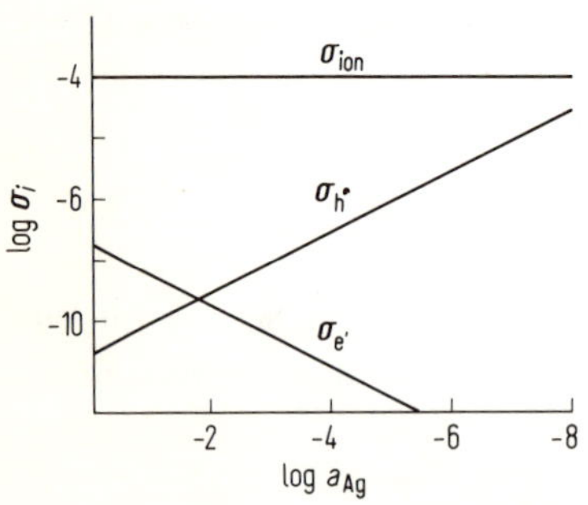

Fig. 4-2. The electrical partial conductivities σ_i for ions and for electronic charge carriers in AgBr at T = 277 °C as a function of the silver activity a_{Ag} according to C. Wagner [8].

In Table 5 are the standard enthalpies and entropies of reactions (4-12), (4-19), and (4-21) according to Kröger [18]. From these values we can calculate the equilibrium constants $K_i = \exp - \Delta G_i^0/RT$, and so all defect concentrations can be calculated as functions of the temperature.

By using eq. (4-16) it can be shown that for $a_{Ag} = 1$ at $T = 277$ °C, δ is about $+10^{-12}$, and for $p_{Br_2} = 1$ atm at $T = 277$ °C, δ is about -10^{-7} [8]. This will be of interest in the later discussion of the elementary photographic process.

Table 5. Entropies and enthalpies for defect reactions in silver bromide according to [18].

	$\Delta S°$ (cal/mole K)	$\Delta H°$ (kcal/mole)
$Ag_{Ag}^x = V'_{Ag} + Ag_i^{\bullet}$ (Frenkel)	25.6	29.3
$Ag_{Ag}^x + Br_{Br}^x = V'_{Ag} + V_{Br}^{\bullet} + AgBr$ (Schottky)	−13.3	36
$0 = e' + h^{\bullet}$	25	78
$1/2\, Br_2 + Ag_{Ag}^x = AgBr + V'_{Ag} + h^{\bullet}$	4.9	25.4

Generally it may be assumed that the standard free energies ΔG_i^0, and thus also ΔH_i^0 and ΔS_i^0, for defect reactions are unique functions of the molar volume [9]. However, the molar volume itself is affected by the lattice expansion with temperature (approximately a linear increase with temperature) and also by the change in volume due to the increasing point defect disorder as the temperature rises. Therefore, the assumption that ΔG_i^0, ΔH_i^0, and ΔS_i^0 are independent of temperature only holds in the first approximation. Actually, near the melting point of AgBr, the increases in lattice expansion, conductivity, etc. with temperature are greater than exponential. This can be partially or totally related back to the change in molar volume with temperature [10]. A number of authors points out the possibility that, in addition to the Frenkel disorder, Schottky disorder also becomes important in AgBr near the melting point [11, 18]. This would likewise explain the fact that the increase of electrical conductivity or of defect concentrations with temperature is greater than exponential.

As yet we have neglected the concentrations of foreign atoms relative to the concentrations of the inherent defects. If small additions of foreign substances are made, within their solubility limits, then the equilibrium constants of the internal and external equilibria as formulated above do not change. However, site balances and electroneutrality conditions must be modified. For example, if $CdBr_2$ is added to AgBr, and minority defects are neglected, then the electroneutrality condition is as follows:

$$(\mathrm{Cd}_{\mathrm{Ag}}^{\bullet}) + (\mathrm{Ag}_{\mathrm{i}}^{\bullet}) - (\mathrm{V}_{\mathrm{Ag}}') = 0 \tag{4-25}$$

Using eq. (4-19), then, we obtain the vacancy concentration $(\mathrm{V}_{\mathrm{Ag}}')$ as a function of the dopant concentration $(\mathrm{Cd}_{\mathrm{Ag}}^{\bullet})$:

$$(\mathrm{V}_{\mathrm{Ag}}') = 1/2\,(\mathrm{Cd}_{\mathrm{Ag}}^{\bullet}) + [1/4\,(\mathrm{Cd}_{\mathrm{Ag}}^{\bullet})^2 + K_{19}]^{1/2} \tag{4-26}$$

For the limiting case $(\mathrm{Cd}_{\mathrm{Ag}}^{\bullet}) \gg K_{19}^{1/2}$ we find that $(\mathrm{V}_{\mathrm{Ag}}') \simeq (\mathrm{Cd}_{\mathrm{Ag}}^{\bullet})$. That is, the vacancy concentration is completely fixed by the addition of $CdBr_2$. This is called the region of exclusively extrinsic disorder, as opposed to the region of intrinsic disorder. In the extrinsic region, those physical properties of the crystal which depend upon the point defect disorder are functions only of the concentration of dopant. However, in deriving eq. (4-26), it has been tacitly assumed that point defects do not form complexes. This assumption, as shown later, must eventually be modified.

From this example, which has been worked out in some detail, for the concentrations of point defects as functions of temperature and component activities in AgBr we can see how one must proceed in general when solving defect problems in binary ionic compounds. The results can be presented graphically as in Fig. 4-2 for AgBr. Kröger [4] has done this for a large number of binary compounds and for a wide variety of disorder types.

4.2.2. Defects in ternary ionic crystals

The concentrations of point defects can also be easily calculated as functions of the independent variables for ternary ionic crystals such as for example spinel phases of the general formula AB_2O_4. The same principles as were used in the previous section are followed. Firstly, a catalogue of all possible disorder types is made. Table 4 can serve as a model for this. Structural elements with negative excess charge are listed as rows, and those with positive excess charge as columns. From such a table it can be seen that a far greater number of possible disorder types exists than is the case for binary ionic crystals, because we now have one more component and a greater variety of crystallographic sites. Thus, in practice, it is usually quite difficult to ascertain the actual disorder type in crystals higher than binary.

In ternary ionic crystals we have at our disposal, as well as P and T, two further independent variables. These may be chosen from a purely practical viewpoint. Accordingly, there are two external equilibrium conditions which must be formulated. The partial pressure of the electronegative component is easy to measure or control experimentally. In the case of the spinel AB_2O_4, this partial pressure would be p_{O_2} which can be controlled either directly or via auxiliary equilibria with H_2/H_2O or CO/CO_2 gas mixtures. Therefore, the activity of the electronegative component is chosen as one independent variable. The corresponding external equilibrium is then formulated in an entirely similar manner as in eq. (4-12).

Because of experimental difficulties when dealing with compounds with narrow ranges of homogeneity, it is expedient to avoid the use of concentrations as independent variables. Instead, we shall choose the activities of AO or B_2O_3 as the other independent variables. For the limiting case in which the spinel coexists with a neighbouring phase, these variables are completely defined. Let us assume, for example, that in the quasi-binary system AO-B_2O_3 there are no other compounds except AB_2O_4. (This, for example, is the case in the system MgO-Al_2O_3.) Then the activity of AO is equal to unity for AB_2O_4 in equilibrium with AO, and equal to $\exp(\Delta G^0_{AB_2O_4}/RT)$ for AB_2O_4 in equilibrium with B_2O_3. $\Delta G^0_{AB_2O_4}$ is the standard free energy of formation of spinel according to the reaction $AO + B_2O_3 = AB_2O_4$. The expression $a_{AO} = \exp(\Delta G^0_{AB_2O_4}/RT)$ for AB_2O_4 in equilibrium with B_2O_3 is derived by applying the equilibrium condition in the form of the mass action law to the spinel formation reaction and by noting further that the equilibrium constant is given by $K = \exp - (\Delta G^0_{AB_2O_4}/RT)$.

In this discussion, we are considering a quasi-binary system AO-B_2O_3, in which electronic point defects are not the majority defect centers (see page 39). Electronic defects are connected with an excess (or a deficiency) of oxygen, and therefore the composition of the compound cannot be located on the quasi-binary line AO $-$ B_2O_3 in case that one kind of electronic defects predominates along with one other majority center. In the quasi-binary system, the second external equilibrium condition can therefore be suitably formulated by means of the following reaction equation:

$$3\,AO\,(g) + V_i^x + 2\,B_i^{\cdot\cdot\cdot} = B_2O_3\,(g) + 3\,A_i^{\cdot\cdot} \tag{4-27}$$

The corresponding equilibrium condition is then:

$$a_{B_2O_3} \cdot a_{AO}^{-3} \cdot (A_i^{\cdot\cdot})^3 \cdot (B_i^{\cdot\cdot\cdot})^{-2} = K_{27} \tag{4-28}$$

The activities of AO and B_2O_3 are coupled in the quasi-binary system through the Gibbs-Duhem equation as follows:

$$a_{AO} \cdot a_{B_2O_3} = \exp(\Delta G^0_{AB_2O_4}/RT) \tag{4-29}$$

The concentrations of $A_i^{\cdot\cdot}$ and $B_i^{\cdot\cdot\cdot}$ depend upon each other through the material balance, the site balance, the electroneutrality condition, and the internal defect equilibria. If the calculations are performed in detail according to the scheme developed in the previous section, then an equation of the following form is obtained for the dependence of the concentration (i) of an arbitrary type of defect i upon the activity a_{AO} at given P, T, and p_{O_2}:

$$(i) = (i)_{a_{AO}=1} \cdot a_{AO}^{n} \tag{4-30}$$

The calculational procedurc automatically gives a value of the exponent n for a given disorder type in a given ternary compound (in this case, AB_2O_4). A typical example is shown graphically in Fig. 4-3 [12, 13]. Of particular note here is the possibility of changing the disorder type through a change in the component activity. However, as can be seen from Fig. 4-3, one majority disorder center will always be preserved if a change in the disorder type occurs because of a change in the component activities.

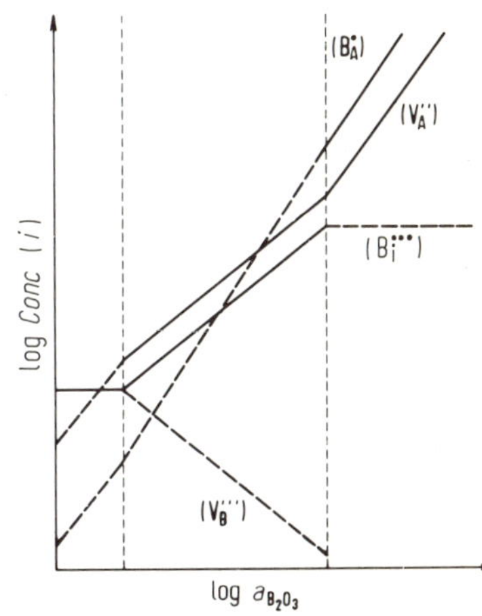

Fig. 4-3. The concentrations of point defects (i) in the ternary crystal AB_2O_4 at constant p_{O_2} as a function of the activity of B_2O_3. The intrinsic disorder is assumed to be Frenkel disorder in the B-ion sublattice: $(B_i^{\cdot\cdot\cdot}) \simeq (V_B''')$.

Finally, several general remarks may be made here. 1. There is no further difficulty involved in including electronic disorder into the calculations. Eq. (4-30) will formally hold for electronic defects as long as the electrons or electron holes obey the laws of ideal solutions (i.e. Boltzmann statistics). Exceptions to this will be discussed later. 2. Since the above discussion is based essentially on thermodynamic arguments, the same equations (4-30) apply for compounds of the same stoichiometry but of different crystal structure, such as Co_2TiO_4 which has the spinel structure, and Co_2SiO_4 which crystallizes in the olivine structure. For compounds with a different stoichiometry than AB_2O_4, such as the compounds A_2BO_4 just mentioned, or the compounds ABO_3 with ilmenite or perovskite structures, the exponents n in eq. (4-30) must be recalculated for each disorder type by following the same sort of procedure [13]. 3. Since ΔG_{AO}^0 and $\Delta G_{B_2O_3}^0$ are generally much more negative than is $\Delta G_{AB_2O_4}^0$, the oxygen activity can be changed over far greater ranges than the activity of AO (or B_2O_3) without that the ternary compound decomposes. Therefore, one may anticipate that the concentration of point defects can be changed far more by changing the oxygen activity than by changing the activity of AO (or B_2O_3).

With this section as a guide, it is possible to calculate the concentrations of point defects as functions of the component activities also for the case of ternary and higher ionic crystals. This then provides a starting point for the quantitative treatment of solid state reactions in which these compounds take part.

4.3. Defect interactions and associates

In order to illustrate the problem, let us take a KCl crystal which has been doped with $CaCl_2$. The concentration of dopant is much greater than the thermal Schottky disorder of the pure crystal. Thus, in the absence of interaction between the defects, the number of dissolved $Ca_K^{\bullet}$ ions is virtually equal to the number of cation vacancies V_K'. However, as described in section 3.2.2, neutral associates will form by virtue of elastic and electrical interactions. Their energy of formation is of the order of 1 kcal/mole. Therefore, at constant P and T, there exists a dynamic equilibrium of the form

$$Ca_K^{\bullet} + V_K' = [Ca, V]^x \tag{4-31}$$

where $[Ca, V]^x$ designates the neutral defect complex. Let β be the degree of association (i.e. the fraction of dopant ions which are found in the complexes). Then, if N_{Ca} is the total mole fraction of dopant, the following equilibrium condition for eq. (4-31) is obtained:

$$\frac{\beta}{(1-\beta)^2} = N_{Ca} K_{31}\,; \quad K_{31} = z_C \exp - \Delta G_{31}^0/RT \tag{4-32}$$

The factor z_C is the number of possible orientations of the complex in the crystal. In the present case, z_C is 12 [6]. Curves showing the degree of association β as a function of the normalized temperature $RT/\Delta G_{31}^0$ are presented in Fig. 4-4.

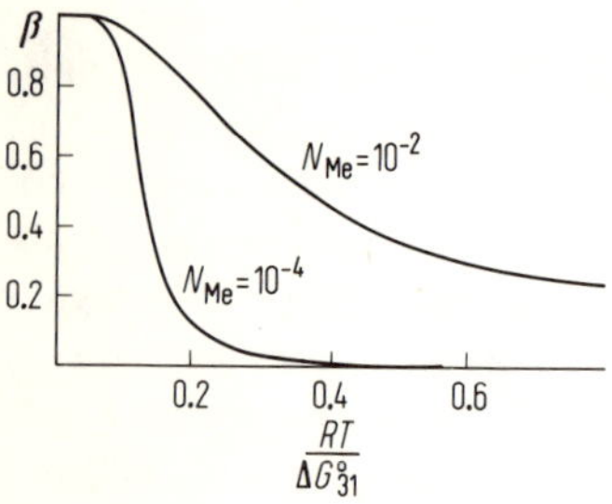

Fig. 4-4. The degree of association β for the complex reaction $Me_C^{\bullet} + V_C' = [Me, V]^x$ (C indicates a cation site) as a function of the normalized temperature for various degrees of doping N_{Me} according to Lidiard [6].

As can be seen, the degree of association β falls off very steeply with increasing temperature for low concentrations of dopant. As well as directly neighbouring defects which form complexes or associates, we must also consider "excited states" of defect associates in which the interacting point defects are more widely separated. K_{31} must then be modified to $\sum_e z_e \exp - \Delta G_{31}^0(e)/RT$. The summation over e indicates a summation over the possible "excited states". It must be noted that, from the electrostatic point of view, a defect complex is a multipole. Therefore, there are interactions, albeit small, between the complexes themselves. At low enough temperatures these interactions lead to aggregates of complexes and finally to precipitation. It follows from this that eq. (4-32) only applies at temperatures which are sufficiently above the solubility line in the phase diagram.

We shall just mention here that the simple association theory may be extended by considering the interactions between defects in solution in a medium with dielectric constant ε [14]. This is analogous to the Debye-Hückel theory of electrolytic solutions. As a result, the mole fractions of charged point defects of sort i in the mass action laws have to be replaced by their corresponding activities which according to Debye-Hückel are of the form

$$a_i = f_i x_i = x_i \cdot \exp - \frac{161.74}{RT} \cdot \frac{z_i^2/\varepsilon}{l(\text{Å}) + a(\text{Å})}$$

f_i is the activity coefficient, z_i the effective charge, l (in Ångström units) is the Debye-Hückel length discussed on page 29, and a (in Angström units) is the distance between defects at which complex formation occurs. In a zeroth approximation, a(Å) can be set equal to the interatomic distance.

These calculations for the formation of associates in KCl doped with heterovalent $CaCl_2$ are illustrations and can be modified without difficulty for other cases of interaction between oppositely charged defects.

4.3.1. Compounds with wide ranges of homogeneity

The defect thermodynamics which we have been discussing up to now applies for the case of very small defect concentrations, such that all interactions between the defects can be neglected in the zeroth approximation. Calculations along the lines of those of Debye and Hückel [14] can be used to treat the interactions in the first approximation if the defect concentrations are small enough. However, the assumptions involved here no longer apply for compounds with wide ranges of homogeneity. Extreme examples would be $Ti_{1-\delta}O$ or $V_{1-\delta}O$ [19]. For the oxides $Mg_{1-\delta}O$ and $Co_{1-\delta}O$, which both crystallize in the rock salt structure, values of δ in air at 1 000 °C range from unmeasurably small values for MgO up to 1% for CoO. For TiO, however, the deviations from the stoichiometric composition lie in the range $+0.56 \geq \delta \geq -0.21$. This is the result of large unequal concentrations of vacancies on the cationic and anionic sublattices [20]. If defects with such large concentrations were randomly distributed over the regular lattice sites, we might then inquire whether the term "crystal structure" could still be clearly defined. Thus, other suggestions have been made as to how the arrangement of defects with such large concentrations in the lattice can be explained. One suggestion is: 1. submicroscopic inhomogeneities such as can be deduced for $Fe_{1-\delta}O$ from the neutron and X-ray diffraction measurements [21, 28]. Here the diffraction effect would seem to be most likely interpreted on the basis of fluctuating clusters of extremely small regions similar to magnetite (Fe_3O_4). Further suggestions are: 2. discrete phases, whose deviations from simple stoichiometric ratios result from the transposition of the structural units of a matrix crystal relative to one another (shear structures) [22]; this shearing is equivalent to an annihilation of point defects (for example anion vacancies) in the shear planes and can occur in one, two, or three dimensions; and 3. discrete phases whose compositions can be formulated as Me_nX_{n+1} with relatively large integral values of n, and which contain superstructures of ordered defects [23].

In the first case mentioned above, if the deviations from stoichiometry become too large (i.e. if the number of statistically distributed defects or defect clusters becomes too great), then the crystal structure must eventually become unstable. In order to better understand the reason for this, we may recall the "titration curve" in Fig. 4-1. Here the logarithm of the relative activity of a component was plotted as a function of the deviation δ from the stoichiometric composition. It was assumed that the free energies of formation of the defects are constant (i.e. independent of the concentrations of point defects). This assumption must, of course, fail for the higher defect concentrations which are being considered in the present section. Detailed calculations have been performed by several authors [24, 25]. These calculations show that, as a result of the interactions between defects, the "titration curves" resemble the well-known isotherms of the van der Waals equation below the critical temperature. Now, the activity of a component in a binary system cannot increase when the concentration of this component decreases. Therefore, in complete analogy to the van der Waals equation, the calculations predict the existence of unstable regions in which the crystal separates into two phases. Beyond these basic calculations whose value is essentially heuristic, there have been no essential improvements in the thermodynamics of solids containing high concentrations of defects [27].

The influence of defect clusters or extended defects in the sense of this section on the mobility of structure elements of a crystal or, more generally, on the reactivity of solids has not yet been explored and is certainly an important field of future research. It is understood, however, that beside shear planes, defect superstructures, or defect clusters, there are also free point defects present in the crystal lattice of compounds with extended ranges of homogeneity.

4.4. Some relations from statistical thermodynamics

The relationships between defect concentrations and the independent thermodynamic variables which were derived in sections 4.1 and 4.2 are based upon the assumption that the chemical potentials of the defects are of the form $\mu_i = \mu_i^0 + RT \ln N_i$, and that the configurational entropy S^C is of the form $-R\sum_i N_i \ln N_i$ where the summation is over all structural elements. This presupposes that Boltzmann statistics can be applied to the structural elements of the crystal, and expresses the assumption that the defect centers obey the laws of ideal dilute solutions.

However, electrons in solids obey Fermi statistics. Now, Boltzmann statistics makes no assertions regarding the number of particles which are permitted in the cells in phase space, but rather, it examines the probability of finding a given number of particles in the individual cells. Fermi statistics, on the other hand, asserts from the start that, because of the Pauli principle, only two particles are permitted in any cell of phase space [7]. In order to show how Fermi statistics affects the formulation of the chemical potentials of electronic defects in crystals, and how the mass action laws are thereby affected, let us take the compound silver sulfide $Ag_{2+\delta}S$ as an example [15].

Silver sulfide is a purely electronic conductor. Hall effect measurements show that conduction is by excess electrons. Above 180 °C (α-Ag_2S), a high degree of cation disorder is indicated by X-ray measurements [16]. Furthermore, the deviation of $Ag_{2+\delta}S$ from the stoichiometric composition as a function of the silver activity is very accurately known. This is shown in Fig. 4-5. δ is the excess of silver above the stoichiometric composition of the sulfide. Since the silver exists in the form of ions, δ is equivalent to the concentration of excess electrons.

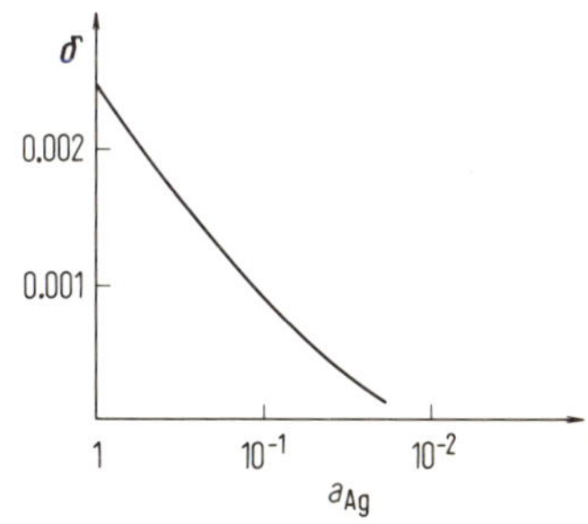

Fig. 4-5. The deviation δ from the stoichiometric composition for silver sulfide as a function of the silver activity at $T = 300$ °C according to Wagner [15].

If these measurements are analyzed by the method of formal mass action laws as in section 4.2, then discrepancies between theory and experiment are found. These discrepancies can be explained if we remember that, at sufficiently high concentrations, the electrons in solids become degenerate. That is, they no longer obey the simple laws of Boltzmann statistics which form the basis of the mass action laws in defect thermodynamics. For $Ag_{2+\delta}S$ this degeneracy must be taken into account because of the high concentration of electrons.

Since in $Ag_{2+\delta}S$ we are dealing with electrons of high mobility, we start out with the band model and consider the electrons to be quasi-free particles in the field of the periodically ordered ions. The influence of the periodic potential upon the mobility of an electron is then taken into account by introducing the so-called "effective mass" m^*, which is different from the rest mass of the electron. Let p be the momentum of an electron, where p vanishes at the lower edge of the band with potential energy E_0. Then the number Z of cells in that sector of phase space with momentum between p and $p + \mathrm{d}p$ calculated per unit volume for the case of an isotropic material is given by:

$$\frac{\mathrm{d}Z}{V} = 2\frac{4\pi p^2\,\mathrm{d}p}{h^3} \tag{4-33}$$

h is Planck's constant, and h^3 is the volume of the elementary cell of the phase space. Because of the spin of the electrons, each cell of the phase space can be occupied by two electrons, and this accounts for the factor 2 in eq. (4-33). If we switch from momentum space to energy space and observe that $p^2/2\,m^* = E - E_0$, we obtain from eq. (4-33):

$$\frac{\mathrm{d}Z}{V} = \frac{4\pi\sqrt{(2m^*)^3\,(E - E_0)}}{h^3}\,\mathrm{d}E \tag{4-34}$$

Eq. (4-34) gives the number of electronic states in the energy interval between E and $E + \mathrm{d}E$ calculated per unit volume. According to Fermi statistics, the distribution function for electrons with energy E is given by [7]

$$f(E) = \left[1 + \exp\frac{E - E_\mathrm{F}}{kT}\right]^{-1} \tag{4-35}$$

where E_F is the energy of the Fermi level. At 0 K, all energy states up to the Fermi level are occupied with two electrons. The total number of electrons per unit volume, $n_{e'}$, is given by multiplying the number of electronic states by the Fermi distribution function and integrating over the entire band:

$$n_{e'} = \int_{E_0}^{\infty} \frac{4\pi\sqrt{(2m^*)^3 (E - E_0)}}{h^3 [1 + \exp(E - E_F)/kT]} \, dE \tag{4-36}$$

If the substitutions $u = (E - E_0)/kT$ and $\mu_{e'} = E_F - E_0$ are made, then the normalized mole fraction of electrons is given from eq. (4-36) as:

$$N_{e'}/D_{e'} = \frac{2}{\pi^{1/2}} \int_0^{\infty} \frac{u^{1/2}}{1 + \exp(u - \mu_{e'}/kT)} \, du \tag{4-37}$$

where

$$D_{e'} = \frac{V_m}{N_0} \cdot 2 \left(\frac{2\pi m^* kT}{h^2} \right)^{3/2} \tag{4-38}$$

$\mu_{e'}$ is the chemical potential of an electron. This can be immediately determined in the limit of very small electron concentrations (i.e. for $N_{e'} \ll D_{e'}$). In this case, the denominator of the integrand must be very large compared to unity. That is, $\exp(u - \mu_{e'}/kT) \gg 1$. For this limiting case, the electron concentration is given by eq. (4-37):

$$N_{e'} = D_{e'} \frac{2}{\pi^{1/2}} \int_0^{\infty} \frac{u^{1/2}}{\exp u} \, du \cdot \exp\frac{\mu_{e'}}{kT} = \text{const} \cdot \exp\frac{\mu_{e'}}{kT} \tag{4-39}$$

By a simple rearrangement of eq. (4-39), the desired expression for the chemical potential $\mu_{e'}$ of the electrons in an ideal dilute solution is obtained in the well-known form $\mu_{e'} = \mu^0_{e'}$

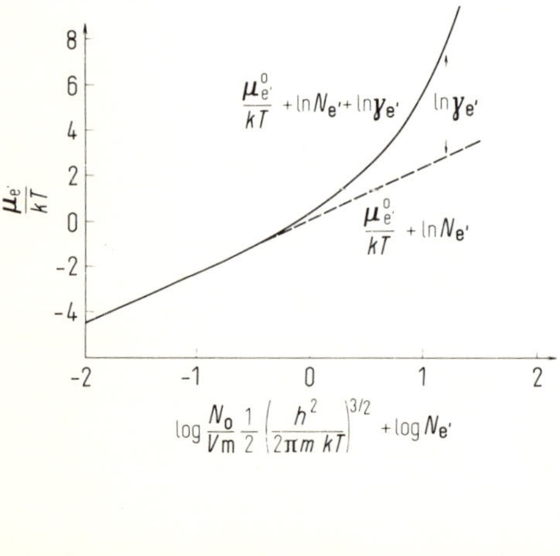

Fig. 4-6. The chemical potential $\mu_{e'}$ of the electron as a function of its concentration in an isotropic crystal. The effective mass is m^*, and $\gamma_{e'}$ is the activity coefficient defined by the equation $\mu_{e'} = \mu^0_{e'} + kT \ln \gamma_{e'} N_{e'}$

$+ kT \ln N_{e'}$. For the numerical evaluation of eq. (4-37), one may refer to the work of J. Mc Dougall and E. C. Stoner [26].

If $N_{e'} > D_{e'}$, then the chemical potential of the electrons may be written as $\mu_{e'} = \mu_{e'}^0 + kT \ln \gamma_{e'} N_{e'}$, where $\gamma_{e'}$ is a formal activity coefficient. $\gamma_{e'}$ can be calculated from eq. (4-37) which is of the form $N_{e'} = F(\mu_{e'}/kT)$ [17]. The quantities are presented graphically in Fig. 4-6 where $\mu_{e'}/kT$ is plotted versus $N_{e'}/D_{e'}$.

From this diagram it can be seen that:

1. If $N_{e'} \ll D_{e'}$, then $\mu_{e'} = \mu_{e'}^0 + kT \ln N_{e'}$. That is, the electron gas obeys the laws of ideal dilute solutions. This implies that $\exp(u - \mu_{e'}/kT) \gg 1$ (i.e. by resubstitution for u and $\mu_{e'}$: $\exp(E - E_F)/kT \gg 1$). Thus, the distribution function in eq. (4-35) can be simplified to the expression $f(E) = \exp - (E - E_F)/kT$. In other words, under these conditions the electrons obey Boltzmann statistics. The band can then be treated as a single energy term E_0, with the density of states given by:

$$\frac{D_{e'} N_0}{V_m} = 2\left(\frac{2\pi m^* kT}{h^2}\right)^{3/2} \tag{4-38 a}$$

2. If $N_{e'} > D_{e'}$ (for Ag_2S at 500 K, $D_{e'}$ is of the order of 10^{-3}), then $\mu_{e'} = \mu_{e'}^0 + kT \ln N_{e'} + kT \ln \gamma_{e'}$, where the logarithm of the activity coefficient $\gamma_{e'}$ can be immediately taken from Fig. 4-6. In this case, the electron gas is said to be partially degenerate. The activity coefficient $\gamma_{e'}$ describes the degeneracy. Deviations from "ideality" ($\gamma_{e'} = 1$) do not arise, as in the case of ionic defects, because of electrical or elastic interactions, but rather they are a result of a maximum permissible occupancy of the existing energy terms.

The methods of this section can be quite generally applied to compounds with relatively high concentrations of electronic defects. This treatment permits us to retain the convenient formal mass action law for defects in solids. This is accomplished by the use of an activity coefficient for electronic defects which can be read off of Fig. 4-6. At the same time, this section has served as a good example of the relationship between statistical calculations and thermodynamic functions. Finally it may be mentioned again that here the chemical potential of electrons and the Fermi energy are synonymous.

4.5. Literature

General Literature:

[1] E. M. Levin, C. R. Robbins, and H. F. McMurdie: Phase Diagrams for Ceramists, The American Ceramic Society, Columbus 1964.
[2] M. Hansen, Constitution of Binary Alloys, 2nd ed., McGraw-Hill Book Comp., Inc., New York 1958.
[3] R. Haase, Thermodynamik der Mischphasen, Springer-Verlag, Berlin 1956.
[4] F. A. Kröger, The Chemistry of Imperfect Crystals, North-Holland Publ. Comp., Amsterdam 1964.
[5] W. Schottky, Halbleiterprobleme IV, Vieweg und Sohn, Braunschweig 1959, p. 235.
[6] A. B. Lidiard in S. Flügge, Handbuch der Physik, Springer-Verlag, Berlin 1957, Vol. XX.
[7] C. Kittel, Introduction to Solid State Physics, 2nd ed., John Wiley and Sons, Inc., New York 1962.

Special Literature:

[8] C. Wagner, Ber. Bunsenges. phys. Chemie *63*, 1027 (1959).
[9] W. Jost, Z. phys. Chemie NF *21*, 202 (1959).

[10] H. Schmalzried, Z. phys. Chemie NF *22*, 199 (1959).
[11] A. W. Lawson, Phys. Rev. *78*, 185 (1950).
[12] H. Schmalzried in H. Reiss, Progress in Solid State Chemistry, Pergamon Press, Oxford 1965, Vol. 2, p. 265.
[13] L. C. Walters and R. E. Grace, J. Phys. Chem. Solids *28*, 239 (1967); *28*, 245 (1967).
[14] E. Hückel, Ergebnisse exakt. Naturw. *3*, 199 (1924).
[15] C. Wagner, J. chem. Physics *21*, 1819 (1953).
[16] P. Rahlfs, Z. phys. Chemie Abt. *B 31*, 157 (1935).
[17] W. Weizel, Lehrbuch der theoretischen Physik, Springer-Verlag, Berlin 1958, Vol. 2, p. 1705 ff.
[18] F. A. Kröger, J. Phys. Chem. Solids *26*, 901 (1965).
[19] N. N. Greenwood, Ionic Crystals, Lattice Defects and Nonstoichiometry, Butterworths, London 1968, p. 131.
[20] M. E. Straumanis and H. W. Li, Z. anorg. allg. Chem. *305*, 143 (1960).
[21] W. L. Roth, Acta Cryst. *13*, 140 (1960).
[22] A. D. Wadsley in L. Mandelcorn, Non-Stoichiometric Compounds, Academic Press, New York 1964, p. 98.
[23] F. Jellinek, Acta Cryst. *10*, 620 (1957).
[24] J. S. Anderson, Proc. Roy. Soc. (London) *A 185*, 69 (1946).
[25] G. G. Libowitz in H. Reiss, Progress in Solid State Chemistry, Pergamon Press, Oxford 1965, Vol. 2, p. 216.
[26] J. McDougall and E. C. Stoner, Phil. Trans. Roy. Soc. *A 237*, 67 (1938).
[27] L. Eyring and M. O'Keeffe, editors, The Chemistry of Extended Defects in Nonmetallic Solids, North-Holland Publ. Comp., Amsterdam 1970.
[28] F. Koch and J. B. Cohen, Acta Cryst. *B 25*, 275 (1969).

5. Chemical diffusion in the solid state

5.1. General remarks

The discussion of a solid state reaction must always be based upon two foundations: upon the atomistic and structural models of mass transport on the one hand, and upon the phenomenological laws of mass transport on the other hand. The most important aspects of the former have been described in the preceding chapters. In the present chapter, a summary of the phenomenological theory of diffusion will be given. Here a distinction must be made between understanding the physical chemistry behind the general phenomenological laws on the one hand and obtaining explicit solutions to the equations in space and time on the other. In this book we shall be mainly concerned with understanding the physical chemistry of diffusional processes in solid phases. For the solution of the equations one should either refer to the excellent monographs on this subject [1, 2, 3, 4] – as far as is possible – or one should accept the advice of a mathematician. In accordance with this expressed purpose, the diffusion equations will at first only be formulated and discussed for one-dimensional problems. In practice, when studying solid state reactions from a fundamental point of view, one should try as far as possible to realize a one-dimensional geometry. There will be cases, of course, where this is not possible because of the nature of the process, as, for example, in the study of powder reactions. The difficulties which arise in such cases will be discussed later.

5.2. Definitions and fundamental relationships

Diffusion means a local non-convective flux of matter under the action of a chemical or – in the case of charged particles – an electrochemical potential gradient. By expressing the flux j_i of particles of type i per unit of the concentration gradient $\mathrm{d}c_i/\mathrm{d}x$ (i.e. by forming the quotient of the measurable parameters j_i and $\mathrm{d}c_i/\mathrm{d}x$), we arrive at a definition of the partial chemical diffusion coefficient of the particles of type i:

$$\tilde{D}_i = -j_i/(\mathrm{d}c_i/\mathrm{d}x) \tag{5-1}$$

The reference system for the measurement of the particle fluxes will be discussed later in detail. The diffusion coefficient has the dimension cm²/sec. The units of j_i can be arbitrarily chosen. However, the units of c_i must then correspond. The following possibilities exist: c_i (mole/cm^3, g/cm^3, equivalents/cm^3); j_i (mole/cm^2 · sec, g/cm^2 · sec, equivalents/cm^2 · sec).

In an isothermal, isobaric, isotropic, and single-phase system of components, the flux equations for neutral particles can be expressed in a linearized general formulation as:

$$\begin{aligned} j_1 &= -L_{11}\,\mathrm{d}\mu_1/\mathrm{d}x - L_{12}\,\mathrm{d}\mu_2/\mathrm{d}x - \cdots L_{1n}\,\mathrm{d}\mu_n/\mathrm{d}x \\ &\vdots \\ j_n &= -L_{n1}\,\mathrm{d}\mu_1/\mathrm{d}x - L_{n2}\,\mathrm{d}\mu_2/\mathrm{d}x - \cdots L_{nn}\,\mathrm{d}\mu_n/\mathrm{d}x \end{aligned} \tag{5-2}$$

For ionic crystals, the chemical potential must be replaced by the electrochemical potential. Eq. (5-2) says that the flux of the particles of type i is also dependent upon the fluxes of all other particles. The coefficients L_{ik} are called transport coefficients. In certain reference systems [6] the Onsager reciprocal relationships $L_{ik} = L_{ki}$ may be applied to these coefficients. On the basis of experience it can be stated that the influence of neutral atomic particle fluxes upon one another in solid phases can usually be neglected. That is, the off-diagonal elements of the L_{ik}-matrix can be set approximately equal to zero. (The complex influence of correlation effects will be treated in detail in later sections.) This follows from the fact that defects exist in very low concentrations and so move independently of each other. And, of course, it is through the motion of defects that the particles are rendered mobile. Then, through simplification of eq. (5-2), we obtain:

$$j_i = -L_{ii}\frac{\mathrm{d}\mu_i}{\mathrm{d}x} = -L_{ii}RT\frac{\mathrm{d}\ln a_i}{\mathrm{d}x} \tag{5-3}$$

For the case of an ideal dilute solution, the partial diffusion coefficient $\tilde{D}_i$ in eq. (5-1) becomes independent of concentration and becomes equal to the component diffusion coefficient D_i of the particles i which will be discussed later. In this case, a comparison of eqs. (5-1) and (5-3) shows that $L_{ii} = D_i c_i / RT$. By substituting this relationship into eq. (5-3) and generalizing, we obtain for the particle flux:

$$j_i = -\frac{D_i c_i}{RT}\frac{\mathrm{d}\mu_i}{\mathrm{d}x} \tag{5-4}$$

The significance of the diffusion coefficient D_i must be discussed separately for different systems such as ideal dilute solutions, phases with wide ranges of homogeneity, and compounds with narrow ranges of homogeneity. The flux equations (5-2) and (5-3) are purely phenomenological in nature. These equations can be interpreted atomistically on the basis of the previously developed models of crystals, and especially upon models of disorder in crystals. Two concepts are of fundamental importance here. These are the mobility b_i and the jump frequency Γ_i of particles of type i.

The particle mobility b_i is defined as the average drift velocity v_i per unit force. In this way, explicit statements about collisions and the microscopic paths of the particles i are avoided. Under the assumption of uncorrelated collisions for particles which are originally in random motion, it can be shown that the application of a force leads to a drift velocity which is proportional to the acting force $\underline{F}$. For the purpose of illustration, consider an ionic crystal to which an electric field is applied. The force $\underline{F}$ acts as $\underline{F} = e_0\mathfrak{E} = -e_0\,\mathrm{d}\phi/\mathrm{d}x$ upon particles which carry unit charge. Then:

$$v = b\underline{F} \tag{5-5}$$

Per unit time,

$$j_i = c_i b_i \underline{F} \tag{5-6}$$

particles of type i pass through a unit area which is perpendicular to the particle flux j_i, where c_i is the local concentration of the particles per unit volume. If $\underline{F}$ is expressed as the gradient of a potential function P (i.e. if $\underline{F} = -\text{grad } P$), then it follows from eq. (5-6) for a one-dimensional case that:

$$j_i = -c_i b_i \frac{\mathrm{d}P}{\mathrm{d}x} \tag{5-7}$$

This equation should be compared to eq. (5-3). For particles i moving under the influence of a chemical potential gradient, eqs. (5-3), (5-4), and (5-7) may be combined to give the so-called Nernst-Einstein equation which relates the mobility b_i to the diffusion coefficient D_i ($k = R/N_0$):

$$D_i = kTb_i \tag{5-8}$$

In order now to relate the average jump frequency Γ_i of a particle i to b_i or to D_i, let us analyze the particle flux passing through an imagined unit interface lying between two lattice planes of an ideal isotropic solution. The planar density (number of atoms per cm^2) of particles of type i is $N_{i,x}$ for the lattice plane in front of the unit interface, and $N_{i,x+a_i}$ for the lattice plane behind the unit interface. The distance between the two planes is a_i. Γ_i is the average jump frequency of the particles i. The flux j_i is then given by combining the partial fluxes in the positive and negative directions. That is:

$$j_i = -g\Gamma_i \, [N_{i,x+a_i} - N_{i,x}] \tag{5-9}$$

where g is a purely geometrical factor which gives the jump probability of the particles in the particular direction under consideration. For the present case of an isotropic cubic lattice, $g = 1/6$. If the planar density N_i is converted to local density c_i according to $c_i = N_i/a_i$, then the concentration gradient is given by $\mathrm{d}c_i/\mathrm{d}x = (N_{i,x+a_i} - N_{i,x})/a_i^2$. From eq. (5-9) it then follows that:

$$j_i = -g\Gamma_i a_i^2 \frac{\mathrm{d}c_i}{\mathrm{d}x} \tag{5-10}$$

By comparing eq. (5-10) with eqs. (5-1) and (5-8) it may be seen that:

$$D_i = kTb_i = g\Gamma_i a_i^2 \tag{5-11}$$

This equation relates the diffusion coefficient D_i, which was defined phenomenologically, to the mobility b_i and to the jump frequency Γ_i of the diffusing particles. It should be stressed here that a jump from an occupied lattice plane to the next plane is only possible if there is an empty site available there. Accordingly, Γ_i is proportional to the concentration of these vacant sites, and therefore in general substantially smaller than the vibrational frequency ν of particles in the lattice.

In systems which contain electrically charged particles with charge z_i, the electrochemical potential of particles of type i is defined in the well-known way as:

$$\eta_i = \left(\frac{\partial G}{\partial n_i}\right)_{P,T,n_j} = \mu_i + z_i F\phi \tag{5-12}$$

where μ_i is the chemical potential, and ϕ is the electrical potential. z_i is the valence (or charge) of the ions of type i. The additivity of the chemical and electrical potentials in eq. (5-12) arises from the fact that in order to create an electrical field of the order of 1 volt/cm at finite distances, charges of the order of $\leq 10^{-16}$ moles are required, and such small quantities are not chemically measurable. In such systems the transport equation for the particles i may be written as

$$j_i = -\frac{D_i c_i}{RT}\frac{\mathrm{d}\eta_i}{\mathrm{d}x} \tag{5-13}$$

which is completely analogous to eq. (5-4).

The average drift velocity b_i per unit force may be replaced by the electrochemical mobility u_i (i.e. by the average drift velocity per unit electric field strength). Then, through the relationship $b_i/u_i = 1/z_i e_0 = N_0/z_i F$, the modified Nernst-Einstein relation is obtained as:

$$D_i = \frac{RT}{z_i F} u_i \tag{5-14}$$

e_0 is the elementary electric charge, and $e_0 N_0 = F$ is the Faraday constant. Furthermore, when the chemical potential of the particles of type i is constant, then the electrical current density I_i and the particle flux j_i are coupled through the relation:

$$I_i = \sigma_i \mathfrak{E} = z_i F j_i = \frac{D_i c_i}{RT}(z_i F)^2 \frac{\mathrm{d}\phi}{\mathrm{d}x} \tag{5-15}$$

$\mathfrak{E} = -\mathrm{d}\phi/\mathrm{d}x$ is the electrical field strength, and σ_i is the partial electrical conductivity of particles of type i. It then follows by direct comparison that:

$$D_i = \sigma_i \frac{RT}{c_i}\frac{1}{(z_i F)^2} \tag{5-16}$$

and:

$$\sigma_i = z_i F c_i u_i \tag{5-17}$$

Eq. (5-13) may then be rearranged to give:

$$j_i = -\frac{\sigma_i}{(z_i F)^2}\frac{\mathrm{d}\eta_i}{\mathrm{d}x} \tag{5-13a}$$

For any specific case, one should use that form of the flux equation which is most appropriate to the particular problem at hand and for which the necessary measured quantities are available.

5.3. Diffusion mechanisms

The elementary steps of diffusion are only possible because of the existence of lattice defects. A single-phase polycrystalline material consists of crystallites containing point defects, dislocations, and low-angle grain boundaries. Between the crystallites there are high-angle grain boundaries and, finally, the solid itself is bounded by surfaces. All these individual imperfections must be treated as different phases when we are considering diffusion in the lattice, since the mobility of the diffusing particles can be very different in the various lattice regions and in the imperfections. In the following, the atomic diffusion mechanisms in each of these various regions will be individually discussed.

5.3.1. Vacancy diffusion

As can be seen from Fig. 5-1, the elementary step of a particle during vacancy diffusion is associated with an elementary step of a vacancy in the opposite direction.

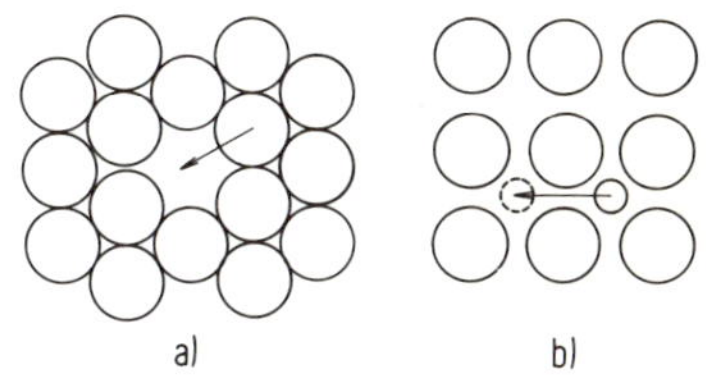

Fig. 5-1. Schematic diagram of the elementary atomic steps during vacancy diffusion (a) and interstitial diffusion (b).

Since the vacancy concentration c_V is generally small in comparison to the particle concentration c_i, the average jump frequencies Γ_i and Γ_V are related as follows:

$$\frac{\Gamma_V}{c_i} = \frac{\Gamma_i}{c_V}; \quad \Gamma_i \simeq N_V \Gamma_V \tag{5-18}$$

N_V is the mole fraction of vacancies V. In this derivation it is assumed that only particles of type i are found on the sublattice under consideration. In words, eq. (5-18) says that, on the average, n particles of type i have each made one jump when the vacancy has made n jumps. Under the aforementioned assumptions, the mobilities and diffusion coefficients are given by eqs. (5-11) and (5-18) as:

$$b_i = N_V b_V \quad \text{and} \quad D_i = N_V D_V \tag{5-19}$$

If there are several kinds of particles in one sublattice, then their individual site-exchange frequencies with the vacancies will in general be different. In this case, the relationships in eq. (5-19) must be modified through the use of a factor f, which is called a correlation factor. This will be discussed more fully in a later section. The calculation of f for the general case mentioned here is extremely difficult.

5.3.2. Interstitial diffusion

In this case, the mobile particles of type i are found on interstitial sites (see Fig. 5-1). Relationships analogous to eqs. (5-18) and (5-19) apply. Vacancy mobilities and vacancy concentrations are simply replaced by interstitial mobilities and interstitial concentrations. In so doing, however, one implicitly assumes that thermodynamic equilibrium is attained between particles of type i on regular lattice sites and those on interstitial sites. This requires that there be a sufficiently rapid exchange of particles between the regular lattice and the interstitial sites. This exchange can occur either by means of a very small concentration of vacancies in the regular lattice, or else it can occur when interstitial particles push particles off of the regular lattice into interstitial sites and take over these regular lattice sites for themselves (see also the next section). The exchange rate can be measured by a defect concentration relaxation. As will be shown later, this rate can also be calculated.

A further variation of the type of diffusion which we are discussing here is interstitial diffusion in solutions in which the solute is found only on the interstitial sites of the solvent matrix (interstitial solutions). For example, in an Fe-C solution, carbon exists in the interstices of the regular matrix of iron atoms (α or γ iron), and the carbon can move only through these interstices.

5.3.3. Interstitialcy diffusion

The mechanism of interstitialcy diffusion is shown in Fig. 5-2. This is distinguishable from interstitial diffusion proper.

Fig. 5-2. Schematic diagram of the elementary atomic step in interstitialcy diffusion.

In this mechanism, an interstitial ion or atom moves onto a regular lattice site by shoving the particle which was originally there onto an interstitial site. If the particles are charged, and if a linear impulse is assumed, then in one elementary step of interstitialcy diffusion the electrical charge is transported twice the distance of each of the two individual ions involved. However, the advancement of the two particles will not necessarily occur in a straight line (i.e. collinear).

5.3.4. Ring mechanism

The simplest elementary diffusional step in a crystal which could be envisioned would be the direct exchange of two neighbouring atoms. However, for any reasonably dense packing, the activation energy for such a process would be very high because of the large distortion of the lattice neighbourhood which would occur during this elementary diffusional step. Therefore, the probability of such a process is very low. However, far less activation energy would be required for the case in which three or more atoms take part in a cyclical exchange. In ionic crystals, such a mechanism could give rise to diffusion, but not to charge transport, since such a cyclical process would not alter the charge configuration. Up until now the existence of such a process has not been experimentally demonstrated.

5.3.5. Diffusion in surfaces, interfaces, and dislocations

For the most part, regions of lattice imperfections in a solid are regions of increased mobility of structural elements. Depending upon the fraction of the total material which is included in the distorted regions, and depending upon the temperature, the diffusion in these regions can exceed the volume diffusion which we have been discussing up to now. This can have a decided effect upon the kinetics of solid state reactions. A rough calculation can serve to make the situation clear. Let the jump frequency Γ_L of particles in a lattice containing point defects be 10^n per second, and let the jump frequency Γ_D of particles in the distorted region of an inner or outer surface or of a dislocation line be 10^{n+m} per second. The diffusion in the distorted region will become comparable in magnitude to the volume diffusion when the ratio R of the number of particles in the distorted regions to the number in the undistorted regions becomes greater than the ratio Γ_L/Γ_D (i.e. when $R > 10^{-m}$). In general, the activation energy for volume diffusion is higher than that for other diffusion mechanisms. Thus, it is to be expected that, in fine-grained samples or in samples with high dislocation densities, diffusion will be structure-sensitive at lower temperatures, while near the melting point, volume diffusion, which is independent of the line- and planar crystal imperfections, will prevail.

Surface diffusion occurs when particles of the crystal or adsorbed foreign particles move along crystal surfaces. Since surfaces usually exist in contact with other phases, specific adsorption and chemisorption effects occur. Because of these effects, and because of the complicated structure of the surface, the diffusional processes here are very dependent upon the experimental conditions. For instance, depending upon the crystal indices of the lattice plane at the surface, the number of kinks and jogs can vary. The concentration of mobile adsorbed particles i is determined by equilibria of the type: i (kink) $\rightleftharpoons i$ (ad). Thus, the surface diffusion coefficient is dependent upon the indices of the lattice plane at the surface. All in all, the large scatter in measured values of surface and interfacial diffusion coefficients is quite understandable. On silver, for example, an increase in the surface diffusion coefficient of silver atoms is observed when oxygen is adsorbed, even though the activation energy for diffusion is thereby also increased. This compels us to assume that the length a_i of the elementary diffusional step for surface diffusion (see eq. (5-11)) increases markedly when oxygen is adsorbed on silver [7].

In and near grain boundaries and phase boundaries one observes a loosening of the crystal. Thus, the average jump frequency Γ_i of particles i is in general greater here than it is in the bulk of the crystal for comparable or higher concentrations of defects. At the same time, the activation energy for diffusion in these interfaces is generally lower than the activation energy for volume diffusion. The activation energy for surface diffusion proper should be closely connected with the enthalpy of vaporization. This is obvious if we consider that when a particle becomes activated in order to make an elementary diffusional step at the surface, it is part way to becoming completely separated from the surface. That is, it is part way to vaporizing. Surface diffusion or interfacial diffusion can be studied in many ways [8]. Examples are: 1. autoradiography, 2. electron beam microanalysis, 3. field ion and field emission microscopy, 4. measurements of the rate of leveling out of grooves in a surface [9], and 5. sectioning techniques. By means of this last technique, for example, a tracer diffusion coefficient of the form $D^*_{Ag} = 0.895 \exp - 45.95/RT$ has been found for polycrystalline silver at high temperatures. Below 700 °C, however, the diffusion coefficient takes the form $D^*_{Ag} = 0.23 \times 10^{-4} \exp - 26.4/RT$. This is indicative of volume diffusion at high temperatures, and grain boundary diffusion at lower temperatures [10]. This supposition is supported by the following argument. The grain size of silver, after annealing at high temperatures, is about 0.1 cm in diameter. The width of the grain boundary can be estimated to be about 3×10^{-8} cm. Therefore, by calculating the ratio of atoms at the interface to atoms in the undistorted crystal, the ratio of jump frequencies Γ_D/Γ_L can be calculated, as discussed above, to be 10^6. This is a quite plausible figure. A phenomenological solution of this diffusion problem is given in section 5.5.5.

Dislocations must also be taken into consideration as possible high mobility paths for particles. The density ϱ_D and the spatial arrangement of dislocations are very difficult to control. Thus, even for single crystals, transport coefficients can be structure-sensitive at temperatures less than about half the melting temperature where volume diffusion no longer predominates.

Two transport phenomena which are connected with dislocations will be mentioned for illustration. 1. During the diffusion of a foreign material in a solid solvent (e.g. the diffusion of phosphorus in silicon or of sulphur in iron), a stress field is built up because of changes in the lattice constant with solute concentration. This in turn can cause dislocations to be formed, and plastic flow is involved. The dislocations are then carried along with the diffusion front. In this way, the diffusion coefficient at the reaction front can become altered. 2. If a crystal is deformed and is then subsequently annealed back to complete recovery, most of the dislocations will be found to be ordered as low-angle grain boundaries. If the dislocations serve as paths of rapid transport, then, based upon the previously described model for low-angle grain boundaries, it is to be expected that the transport coefficients will depend upon the angle between the transport direction and the low-angle grain boundary. A strong anisotropy should thus be observed, depending upon whether material is transported in the direction of the low-angle grain boundary or perpendicular to this direction. All these predictions can be experimentally tested [11, 12].

5.3.6. Temperature dependence of diffusion coefficients

From diffusion experiments one learns that the temperature dependence of the diffusion coefficient D_i can be written as $D_i = D_i^0 \exp(-Q/RT)$, which is an Arrhenius-type relation. For example, in the case of carbon diffusion in iron or sodium diffusion in β-alumina an Arrhenius-type relation is obeyed over many orders of magnitude of the diffusivities. According to eq. (5-19), D_i is the product of a point defect concentration and a diffusion coefficient of the point defects. Eq. (4-3) shows that the mole fraction of point defects is an exponential function of the reciprocal of the absolute temperature for low defect concentrations. Therefore, the individual jump frequency of a point defect, or of an atom or ion which is moved by the jump of a point defect, depends upon the temperature as $\exp(-Q'/RT)$.

There are two different ways to explain the experimentally observed temperature dependence. The first method employs the concept of the activated transition state in the absolute reaction rate theory [33]. Here it is assumed that the diffusion particle crosses an activation energy barrier between two equivalent lattice sites. One calculates the probability of the particle being on the saddle-point (transition state) and its velocity there. This implies that an equilibrium distribution of diffusing particles between normal lattice sites and the saddle-points exists. It is further assumed that the diffusing particles in the saddle-point configuration possess only one translational degree of freedom. Then it can be shown that the individual jump frequencies, the mobilities, and the diffusivities of the atoms or ions of type i depend exponentially upon the reciprocal of the temperature, which agrees with the experimental observations. A preexponential factor of the form (kT/h), which has the dimensions of a frequency, is obtained with this theory [34]. It is obvious that the theory outlined here is analogous to Eyring's [32] theory of absolute reaction rates for chemical reactions. The degree to which this theory correctly describes the process of diffusion in solids is open to discussion.

Nowadays it seems more realistic to explain the Arrhenius-type of temperature dependence of atomic jumps in diffusion processes with the help of lattice dynamics. The task is to calculate the probability of a specific configuration in which the neighbours of a diffusing particle in the vibrating lattice have opened the door for a jump. This means that the vibration of the neighbours under consideration is such that a configuration occurs in which there is no resistance for the diffusing particle to pass into the next equivalent lattice position. In a first approximation, the probability of these configurations, and thus, the mobility and the diffusivity of atoms or ions in the lattice, again depend exponentially on the reciprocal of the absolute temperature. The actual calculations are very cumbersome because the whole spectrum of lattice vibrations must be employed. The interested reader is referred to the original paper on this subject [35].

5.4. Correlation effects

The diffusion coefficient D_i was introduced in eqs. (5-4) and (5-13). From eq. (5-11) it can be seen that D_i is a measure of the mobility or of the jump frequency of the particles of type i. Ever since sufficient quantities of stable or radioactive isotopes for most elements have become available, the so-called tracer method of measuring diffusion coefficients has been widely used. In this method, small quantities of isotope are permitted to diffuse into the system under investigation, and isotopic effects are neglected. In completely homogeneous material, the mean square displacement $\overline{x_i^2}$ of the tracer atoms is experimentally determined. The following formula then applies [13]:

$$\overline{x_i^2} = 2\,D_i^*\,t \tag{5-20}$$

where t is the diffusion time. By using this relationship, so-called tracer diffusion coefficients D_i^* can be calculated. The earliest investigators to use the tracer diffusion method set the diffusion coefficients in eq. (5-20) equal to the diffusion coefficients defined by eq. (5-11). These were then called self-diffusion coefficients. However, a detailed analysis of diffusion mechanisms in crystals was later developed [14] which showed that this equivalency of the two diffusion coefficients is only true when diffusion occurs by uncorrelated elementary steps. The case of interstitial diffusion of particles A in an elemental crystal of A fulfills this criterion of uncorrelated elementary steps of the diffusing particles A, because after every successful diffusional step in the interstices, the diffusing particle finds itself in the same geometrical situation as before the diffusional step took place. However, for diffusion via a vacancy mechanism, for example, this is no longer true. Suppose that a tracer atom A* has exchanged places with a vacancy, and has thereby executed an elementary diffusional step. The probability that the atom will now exchange sites again with the same vacancy, and thereby cancel the effect of the original exchange, is much greater than the probability that the atom will move in any other direction with the help of another vacancy coming along. For the case of a vacancy diffusion mechanism we say that the elementary steps of the tracer atoms are correlated. The mean square displacement no longer provides a measure of the jump frequency Γ_A of the particles A directly from eqs. (5-11) and (5-20). Rather, an "effective" jump frequency ($f\Gamma_A$) is measured in this way, and f is called a correlation factor.

For purposes of illustration, a few situations will be briefly discussed.

1. Correlation effects during vacancy diffusion in a metal A. As has already been shown, if the vacancy and the tracer atom exchange sites twice in succession, no net motion of the tracer atom takes place. We would like to be able to use the measured displacement of the tracer atom in order to calculate the jump frequency or the mobility of the vacancy. The jump frequency could then be used to calculate the diffusion coefficient in the metal A. In order to do this, then, we must calculate the probability of a repeated exchange of sites between a tracer and the same vacancy. The calculation is based upon the known geometry of the lattice and upon the assumption that the direction of a vacancy jump is independent of the direction of all previous jumps. This can be expressed quantitatively as follows: Let $\overline{x_n^2}\,(A^*)$ be the mean square displacement of the tracer A* and let $\overline{x_n^2}\,(V)$ be the mean square displacement of the vacancy V after n jumps ($\lim n \to \infty$). The correlation factor is then defined, by means of eqs. (5-11), (5-19), and (5-20), as:

$$f = \frac{\overline{x_n^2}\,(A^*)}{\overline{x_n^2}\,(V)} = \frac{D_A^* t_n\,(A^*)}{D_V t_n\,(V)} = \frac{D_A^* \cdot (n/\Gamma_V)\,(c_A/c_V)}{D_V \cdot (n/\Gamma_V)} \simeq \frac{D_A^*}{D_V N_V} = \frac{D_A^*}{D_A} \tag{5-21}$$

where $t_n\,(A^*)$ and $t_n\,(V)$ are the times required by the tracer or by the vacancy to make n elementary steps, and Γ_V is the mean jump frequency of the vacancy V. In the third term of this derivation, the jump frequencies of atom A and tracer A* have been taken to be equal. This is not completely true. In higher approximations this can be taken into account by the so-called isotope effect.

In the zeroth approximation, the correlation factor is given by $f = 1 - 2/z$, where z is the number of nearest neighbours in the lattice in which the particles are diffusing [4]. The derivation of this approximation is given later. An exact calculation of f is performed as follows. All individual steps x_n are taken to be lattice vectors of equal length a. These are squared and added to give the mean square displacement:

$$\overline{x_n^2}\,(A^*) = na^2 \left[1 + \frac{2}{n} \overline{\sum_{k=1}^{n-1} \sum_{j=1}^{n-k} \cos \Theta_{j,j+k}} \right] \tag{5-22}$$

$\Theta_{j,j+k}$ is the angle between the lattice vectors of step j and step $j+k$. If all directions for a step are equally probable – that is, if the diffusion is uncorrelated – then the mean of the double summation vanishes, and we are left with the well-known equation for the mean square displacement of particles diffusing by uncorrelated jumps in an isotropic medium: $\overline{x_n^2} = na^2$. This relationship also follows from eqs. (5-11) and (5-20). Consequently, the correlation factor f is given as:

$$f = \left(1 + \frac{2}{n} \overline{\sum_{k=1}^{n-1} \sum_{j=1}^{n-k} \cos \Theta_{j,j+k}} \right)_{n \to \infty} \tag{5-23}$$

For further mathematical treatment of this expression, the reader may refer to the special literature [5, 15]. For cubic systems in which only one simple diffusion mechanism is present, eq. (5-23) can be simplified to:

$$f = \frac{1 + \overline{\cos \Theta}}{1 - \overline{\cos \Theta}} \tag{5-24}$$

where $\overline{\cos \Theta}$ is the mean cosine of the angle between two successive jumps of the diffusing particles. In Table 6 is a summary of correlation factors f for the case of vacancy diffusion in crystals of elements with various lattice structures [16].

Table 6. Correlation factors for vacancy diffusion.

Structure	f
Diamond	0.5
simple cubic	0.653
body-centered-cubic	0.727
face-centered-cubic	0.781
hexagonal-close-packed	0.781

2. Correlation effects in a stoichiometric compound with a very narrow range of stoichiometry. For self-diffusion in elemental crystals as discussed under 1., there is no chemical transport. In stoichiometric compounds, however, particle fluxes as defined in eqs. (5-4) and (5-13) occur whenever a chemical potential gradient exists. Such would be the case, for example, during tarnishing reactions or during the formation of intermetallic compounds or higher ionic compounds. The diffusion coefficient D_i in eqs. (5-4) or (5-13) was introduced by C. Wagner [17] for this set of reactions as a "component diffusion coefficient". Under the given assumptions, the same conditions occur in the individual sublattices of the compound as occurred in the elemental crystal discussed under 1. Therefore, the equations derived previously may also be applied to give the relationship between the component diffusion coefficients and tracer diffusion coefficients which have been measured in the absence of a chemical potential gradient. In particular, the relationship $D_i^* = fD_i$ from eq. (5-21) still applies. Finally, it should be noted that D_i and D_i^* depend not only upon P and T, as in the case of elemental crystals, but they are also dependent upon the component activities, inasmuch as these activities determine the concentrations of defects, and thus the diffusivities.

3. In a dilute solid solution (cubic system) the mean value $\overline{\cos \Theta}$ may be calculated for the solute B in the solvent A, and then the correlation factor f_B can be calculated from eq. (5-24). In order to perform this calculation, let us define p_r as the probability that, after exchanging sites with a vacancy, the particle B will immediately jump back into the vacancy, thereby returning to its original site. If z is the number of nearest neighbours to the vacancy, then:

$$p_r = \frac{\Gamma_B}{\Gamma_B + (z - 1)\,\Gamma_A} \tag{5-25}$$

If, immediately after an exchange of sites with a particle, a vacancy were only permitted to return to its original site on its very next step (with probability p_r), but were not permitted to return if it had made any further steps, then Θ would be 180°, and the averaged value $\overline{\cos\Theta} = -p_r$. If p_r is reintroduced into eq. (5-24), then it follows that $f \simeq (1 - p_r)^2 \simeq 1 - 2p_r = 1 - 2/z$, which is the value of f given before as a zeroth approximation. Actually, the vacancy does not move away from atom B with the frequency $(z - 1)\,\Gamma_A$. Rather, after further jumps it might remain as a nearest neighbour of B, or it could return to B with a calculable probability, thereby either totally or partially cancelling out the effect of the original jump. As in eq. (5-25) we may write:

$$\overline{\cos\Theta} = \frac{\Gamma_B}{\Gamma_B + Q\Gamma_A} \tag{5-26}$$

where, however, $Q \neq (z - 1)$. Q can be calculated from a detailed analysis of the elementary steps [4]. For the face-centered-cubic lattice, Q has the value 7.15. By substituting eq. (5-26) into eq. (5-24), the correlation factor f_B for a dilute solution of B in A is obtained as:

$$f_B = \frac{Q\Gamma_A}{2\,\Gamma_B + Q\Gamma_A} \tag{5-27}$$

The following limiting cases should be mentioned: 1. If $\Gamma_A \gg \Gamma_B$, then $f_B = 1$, and the particle B diffuses without correlation through A. 2. If $\Gamma_A \ll \Gamma_B$, then f_B is very small. This means that the particle B exchanges places back and forth with a neighbouring vacancy very many times before the vacancy exchanges with an A particle and the elementary step of the B particle becomes effective.

Correlation effects become very hard to visualize for the case of non-dilute solid solutions such as mixed crystals with wide ranges of homogeneity. This is especially true when Γ_A and Γ_B are very different from one another. Quantitative treatments have been given by Manning [5]. Results have also been obtained for the case of ionic crystals [18]. We shall only make reference to these works here. Furthermore, Condit [26] has pointed out that the calculation of correlation factors can become even more complicated for cases in which the diffusing particles cannot move from a regular lattice site to a vacancy in one elementary step, but must rather, so to say, first stop off at one or more interstitial sites. This situation is frequently encountered in crystals with complicated structures and with less dense packing. A complete theoretical analysis of this situation is still lacking.

Finally, it should be mentioned that the calculation of correlation effects for crystals in which fluxes occur as the result of chemical potential gradients yields extra terms not yet pointed out because of the spatial inhomogeneity of defects. This higher approximation has been calculated for the case of a vacancy diffusion coefficient, and is known as a result of the "vacancy wind". It can be neglected in first order calculations [5].

In order to obtain correlation factors experimentally, one may for example compare the component diffusion coefficient which has been measured in flux experiments (when a gradient of the electrochemical potential was acting upon the diffusing particles) with the tracer diffusion coefficient. One also may use the so-called isotope effect: Since two isotopes of the same element in a lattice have different vibrational frequencies and accordingly different jump frequencies, their correlation effects and their diffusivities differ also. One can show that the relative change in D_i of two isotopes is proportional to the correlation factor f. The factor of proportionality, in addition, can be evaluated independently if the isotope masses and the kinetic energies of the vibrational modes around the defect are known [5].

5.5. Steady state and non-steady state diffusional processes

5.5.1. Phenomenological treatment of steady state diffusional processes

In order to demonstrate the principles, the following discussion will be limited to one-dimensional problems (planar diffusion) in isotropic media. For systems with constant electrical potential, the flux equation (5-4) may be written in explicit form as:

$$j_i = -\frac{D_i c_i}{RT}\frac{\mathrm{d}\mu_i}{\mathrm{d}x} = -D_i c_i \frac{\mathrm{d}\ln\gamma_i c_i}{\mathrm{d}x} = -D_i\left[1 + \frac{\mathrm{d}\ln\gamma_i}{\mathrm{d}\ln c_i}\right]\frac{\mathrm{d}c_i}{\mathrm{d}x} \tag{5-28}$$

where γ_i is the activity coefficient, and where the flux is measured relative to the crystal lattice of the solid (lattice reference system). For ideal dilute solutions with constant activity coefficients γ_i, eq. (5-28) reduces to Fick's first law:

$$j_i = -D_i\frac{\mathrm{d}c_i}{\mathrm{d}x} \tag{5-29}$$

The condition of steady state diffusion means that concentrations and fluxes at every point on the x-axis are constant with time. It then follows from eq. (5-29) for the case of a constant diffusion coefficient that the concentration gradient is locally constant everywhere, i.e. independent of x. In other words, there will be a linear concentration profile. Examples of such a steady state diffusional process would be the diffusion of numerous gases through metal foils when constant but different partial gas pressures are maintained on either side of the foils. In the system palladium-hydrogen, because of the high diffusivity of hydrogen, direct use is made of the diffusion of the gas through the metal for purposes of gas purification.

If the diffusion coefficient D_i is dependent upon concentration, then the condition of steady state diffusion requires that the concentration gradient be large in regions where the diffusion coefficient is low (i.e. in regions of high resistance to diffusion) in order that the flux remains locally constant everywhere. This can be seen from eq. (5-29). In regions where the diffusion coefficient is large, on the other hand, the concentration gradient will be low. An example is presented schematically in Fig. 5-3. For the diffusion of carbon through iron foil the inverse situation applies [27].

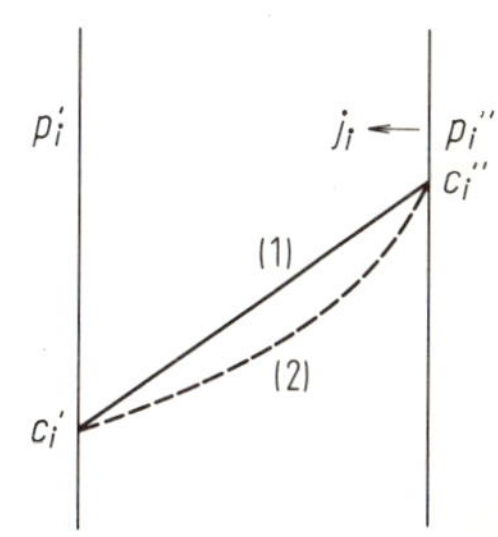

Fig. 5-3. Concentration profiles for stationary diffusional processes through foils. (1) The diffusion coefficient D_i of the solute is independent of concentration. (2) D_i is dependent upon concentration and decreases as the concentration increases.

No further difficulties arise when we come to discuss steady state diffusion in a system in which the activity coefficient γ_i is dependent upon concentration. In such a system, D_i (1 +

$+\mathrm{d}\ln\gamma_i/\mathrm{d}\ln c_i$) is a unique function of the composition. Once again, the concentration gradient will locally assume unique values in order that j_i in eq. (5-28) remains constant in the steady state case.

We shall frequently be examining reactions under conditions which are designated as "quasi-steady state". A "quasi-steady state" process is one which proceeds such that during a specific time interval the flux j_i can be considered to be locally constant, or in which the concentration gradient $\mathrm{d}c_i/\mathrm{d}x$ at a specific point can be considered to be constant in this time interval.

Finally, mention should be made of steady state processes in which a constant diffusional cross-section, such as occurs in the case of one-dimensional diffusion, is no longer present. In these cases, the concentration gradient is not spatially constant when the fluxes and concentrations are constant with time, even for ideal dilute solutions with constant diffusion coefficients. A detailed treatment of such cases will not be given here. Examples for spherical or cylindrical symmetry can easily be found in the literature [1].

5.5.2. Phenomenological treatment of non-steady state diffusional processes in binary systems

Eq. (5-28) is not suitable for describing spatially and temporally variable diffusion processes. Because of the principle of the conservation of mass, however, the divergence of the flux can always be set equal to the time derivative of the local concentration. This leads to Fick's second law which may be written as follows for the case of binary systems with constant diffusion coefficients:

$$\frac{\partial c_i}{\partial t} = \tilde{D}\frac{\partial^2 c_i}{\partial x^2} \tag{5-30}$$

In the form of eq. (5-30), Fick's second law applies only to one-dimensional problems in an isotropic medium. The index i on the diffusion coefficient has been removed in order to make it clear that this is no longer the component diffusion coefficient D_i, but rather, it is the chemical interdiffusion coefficient. Normally, the chemical interdiffusion coefficient will be a function of the individual component diffusion coefficients D_i because of the coupling of the fluxes in the lattice system. When local thermodynamic equilibrium prevails, the coefficients D_i are, in turn, unique functions of the composition. From the thermodynamics of irreversible processes it can be shown [6] that in binary systems there is only one independent transport coefficient, and in general, in n-component systems there can only be $(n - 1)\,n/2$ independent transport coefficients.

If the chemical diffusion coefficient is concentration dependent, then, instead of eq. (5-30) we must write:

$$\frac{\partial c_i}{\partial t} = \frac{\partial}{\partial x}\left(\tilde{D}\frac{\partial c_i}{\partial x}\right) \tag{5-31}$$

The diffusion problem as formulated in eq. (5-31) can be discussed in two ways: 1. A solution of the second-order partial differential equation is sought for a given set of initial and boundary conditions under the assumption that the chemical interdiffusion coefficient $\tilde{D}$ is known as a function of the concentration. This assumption implies that the diffusion mechanism, the coupling of the individual fluxes in the lattice system, and the defect thermodynamics are all known. 2. With the aid of known solutions of Fick's second law, the chemical diffusion coefficient $\tilde{D}$, as well as its concentration dependence, are determined from the results of experimental measurements. The diffusion coefficients so obtained are then interpreted by way of the thermodynamics of the system and the coefficients D_i on the basis of a specific disorder model through the use of defect thermodynamics. (The terms chemical diffusion coefficient and interdiffusion coefficient are used synonymic.)

5.5.3. The chemical diffusion coefficient and its derivation for special cases

The chemical diffusion coefficient $\tilde{D}$ is necessary and sufficient for a phenomenological description of binary diffusion. For one-dimensional diffusion, in which a constant diffusional cross-section is assumed, the average particle velocities v_i ($i = 1, 2$) in an arbitrary coordinate system are given as:

$$v_i = j_i/c_i \quad \text{where } c_i = N_i/V_m \tag{5-32}$$

N_i is the mole fraction of particles of type i, and V_m is the molar volume. In the general case, V_m is dependent upon the mole fraction. The chemical diffusion coefficient must be independent of the chosen coordinate system. The quantity $(v_1 - v_2)$, which is the average velocity of particles of type 1 relative to the velocity of particles of type 2, is also independent of the coordinate system. Therefore, this quantity may be used to give a suitable definition of $\tilde{D}$. According to eq. (5-32), v_i is proportional to j_i. Therefore, by Fick's first law, v_i is also proportional to the concentration gradient. The first step, then, is to divide $(v_1 - v_2)$ by $\partial N_2/\partial x$. Finally, in order to make $\tilde{D}$ conform to the diffusion coefficient in Fick's laws, it is necessary to multiply $(v_1 - v_2)$ by $N_1 \cdot N_2$, so that the definition of $\tilde{D}$ becomes [19]:

$$\tilde{D} = \frac{N_1 N_2 (v_1 - v_2)}{(\partial N_2/\partial x)} = \frac{V_m (N_2 j_1 - N_1 j_2)}{(\partial N_2/\partial x)} \tag{5-33}$$

The coordinate x in eq. (5-33) appears only as a differential, so that the choice of the origin of the coordinate system for this definition is unimportant. Also, no assumption of a constant molar volume has been made. For ideal dilute solutions, $N_1 \rightarrow 1$, $N_2 \rightarrow 0$, and V_m is constant. It then follows from eq. (5-33) that j_2 is equal to $-\tilde{D} \cdot \partial c_2/\partial x$. This is in agreement with eq. (5-29). It can be seen that for ideal dilute solutions the component diffusion coefficient D_i and the chemical diffusion coefficient $\tilde{D}$ are identical.

An expression for the chemical diffusion coefficient for binary diffusion by means of vacancies can be derived in a straightforward way. Assume that the molar volume is independent of concentration. Since transport occurs via vacancies, there will be a flux of vacancies in addition to the fluxes of the components 1 and 2 in the lattice system. If the jump frequency Γ of particles of type 1 into the vacancies is greater than that of particles of type 2, then a local flux of vacancies will occur towards the region of higher concentration of component 1. Under the assumption of internal thermodynamic equilibrium, these vacancies are removed from the crystal at sites of repeatable growth (i.e. dislocations, grain boundaries). Because of this flux and removal of vacancies, the lattice planes in the diffusion region advance. This can be described by a lattice velocity v_L. An observer outside the diffusion zone will see the following fluxes.

$$j_1 = -\frac{c_1 D_1}{RT}\frac{\mathrm{d}\mu_1}{\mathrm{d}x} + v_L c_1$$
$$j_2 = -\frac{c_2 D_2}{RT}\frac{\mathrm{d}\mu_2}{\mathrm{d}x} + v_L c_2 \tag{5-34}$$

Formulating the flux equations in terms of the chemical potential gradient only implies that the fluxes occur in metals or semi-conductors without the build-up of an electrical diffusion potential.

Since the total number of atoms in the crystal remains constant, even though lattice sites are being locally created or destroyed, we may set $j_1 + j_2 = 0$. By using the Gibbs-Duhem relation $c_1 \mathrm{d}\mu_1 + c_2 \mathrm{d}\mu_2 = 0$, we obtain the following expression for the local lattice velocity:

$$v_L = -(D_2 - D_1)\frac{N_1}{RT}\frac{\mathrm{d}\mu_1}{\mathrm{d}x} \tag{5-35}$$

Substitution into eq. (5-34) yields:

$$j_1 = -j_2 = -(N_2 D_1 + N_1 D_2)\left(1 + \frac{\mathrm{d}\ln\gamma_1}{\mathrm{d}\ln c_1}\right)\frac{\mathrm{d}c_1}{\mathrm{d}x} \tag{5-36}$$

By comparison with eq. (5-33), the following expression for the chemical diffusion coefficient is obtained in the present case:

$$\tilde{D} = (N_2 D_1 + N_1 D_2)\left(1 + \frac{\mathrm{d}\ln\gamma_1}{\mathrm{d}\ln c_1}\right) \tag{5-37}$$

Eq. (5-37) was originally derived by Darken [20].

To conclude this section, the chemical diffusion coefficient for the interdiffusion of $SrCl_2$ and KCl will be derived. This will provide an example from the area of diffusion in ionic crystals. At 550 °C, $SrCl_2$ is about 1% soluble in KCl. Assuming negligible thermal disorder, we note that the number of cation vacancies is equal to the number of dissolved strontium ions (i.e. $(V_K') = (Sr_K^{\bullet})$), provided that no association takes place. Diffusion of strontium occurs via cation vacancies. The anion sublattice takes no part in the diffusional processes itself, and so we can use the anion lattice as a natural reference system. Since vacancies and dissolved strontium ions are electrically charged relative to the undisturbed crystal, we must use the gradient of the electrochemical potential as the driving force for diffusion. The fluxes of defects V_K' and of strontium ions $Sr_K^{\bullet}$ relative to the anion sublattice are then given as:

$$j_{V_K'} = -\frac{c_{V_K'} D_{V_K'}}{RT}\cdot\frac{\mathrm{d}\eta_{V_K'}}{\mathrm{d}x}$$
$$j_{Sr_K^{\bullet}} = -\frac{c_{Sr_K^{\bullet}} D_{Sr_K^{\bullet}}}{RT}\cdot\frac{\mathrm{d}\eta_{Sr_K^{\bullet}}}{\mathrm{d}x} \tag{5-38}$$

During the diffusional process, electrical neutrality is maintained in every volume element. Because of this condition, the particle currents will be coupled. That is, $j_{V'_K} = j_{Sr^\bullet_K}$, since V'_K and $Sr^\bullet_K$ carry electrical charges of opposite sign. By means of this condition of electrical neutrality, the unknown differential of the electrical potential ϕ in eq. (5-38) can be eliminated. Since we are concerned with only small concentrations of V'_K as well as of $Sr^\bullet_K$, the laws of dilute solutions apply, and μ_i can be set equal to $\mu_i^\circ + RT \ln c_i$ ($i \equiv$ point defect). Under these assumptions, a simple rearrangement of eq. (5-38) gives:

$$j_{V'_K} = j_{Sr^\bullet_K} = -\frac{2\,D_{V'_K}\,D_{Sr^\bullet_K}}{D_{V'_K} + D_{Sr^\bullet_K}} \cdot \frac{\mathrm{d}c_{Sr^\bullet_K}}{\mathrm{d}x} \tag{5-39}$$

In this case, then, the chemical diffusion coefficient $\tilde{D}$ is given by $\tilde{D} = 2\,D_{V'_K}\,D_{Sr^\bullet_K}/(D_{V'_K} + D_{Sr^\bullet_K})$. Furthermore, the dissolved strontium ions can only diffuse via cation vacancies, and so, in the absence of association with vacancies, $D_{Sr^\bullet_K}$ can be assumed to be small compared to $D_{V'_K}$. Therefore, $\tilde{D} \simeq 2\,D_{Sr^\bullet_K}$. The same result could be obtained by using the definition of the chemical diffusion coefficient from eq. (5-33) as well as eq. (5-38) and the condition of electrical neutrality.

It has thus been shown by means of two examples how the phenomenological laws of diffusion are formulated and what procedures must be followed in special cases in order to express the chemical interdiffusion coefficient in terms of the component diffusion coefficients or, as in eq. (5-39), in terms of the diffusion coefficients of point defects.

In the following section, some important solutions of Fick's second law are presented.

5.5.4. Special solutions of Fick's second law

Along with heat conduction, electrical conduction, and momentum transport, diffusional processes belong to the large family of statistical processes in atomic systems. Since the differential equations of all these processes are of one type, we have at our disposal a large number of solutions for many different initial and boundary conditions. If we wish to study reaction mechanisms and the elementary steps of diffusion, then we should work with the simplest possible initial and boundary conditions in order to give the simplest and clearest mathematical solution, so that we can then devote all our attention to the physico-chemical problem. This can be achieved, among other means, by choosing systems so large that their geometrical boundaries lie outside the diffusion zone. In such cases we have a "semi-infinite" medium. The condition for a medium to be "semi-infinite" for diffusional processes is that the characteristic index $l/\sqrt{\tilde{D}t}$ does not fall below a certain value which can be estimated for the given experimental configuration. l is the characteristic length of the particular system. Furthermore, for concentration dependent (and activity dependent) diffusion coefficients, the concentration (and activity) differences in the reacting system should be kept small enough so that a constant average $\tilde{D}$ can be assumed in the interval under study. In the following discussion, we shall assume a constant chemical diffusion coefficient.

As the first special solution of Fick's second law, let us consider briefly the solution for a so-called point source. By this we mean the following (see Fig. 5-4):

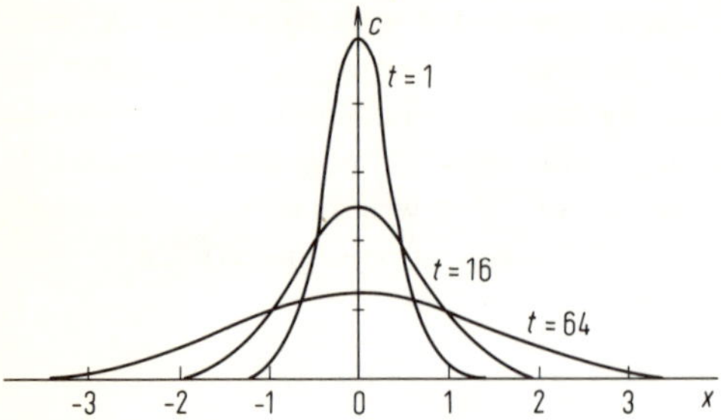

Fig. 5-4. Graphical representation of the solution of Fick's second law for a point source at time $t = 0$. The curves are concentration profiles after various times. The units have been chosen arbitrarily.

At time $t = 0$, an amount s of diffusing substance is placed on a plane of unit area at $x = 0$. Since this amount s does not change during diffusion, it follows that $s = \int_{-\infty}^{+\infty} c \,\mathrm{d}x$ for all times. As long as the boundaries of the system lie outside the diffusion zone, it can be easily confirmed by means of substitution that the following expression satisfies not only Fick's second law (eq. (5-30)), but it also satisfies the initial and boundary conditions. This expression gives the concentration c as a function of position x and time t:

$$c\,(x, t) = \frac{s}{2\sqrt{\pi\,\tilde{D}\,t}} \exp\,(-\,x^2/4\,\tilde{D}\,t) \tag{5-40}$$

The exponential expression has the form of the Gaussian normal distribution curve. If an amount s of diffusing substance diffuses only into one side of a semi-infinite medium – that is, if the diffusing material is placed on the end of a rod with all the other above conditions applying – then the solution (5-40) needs only be multiplied by a factor 2. (The superposition principle for solutions of linear differential equations has been used here.)

Other special solutions are easily obtained through use of the superposition principle by integration of the solution for a point source. Assume that the initial concentration distribution is a uniform concentration $c = c_0$ for $x = -\infty$ to $x = 0$, and $c = 0$ for $x > 0$ as shown in Fig. 5-5 (step function).

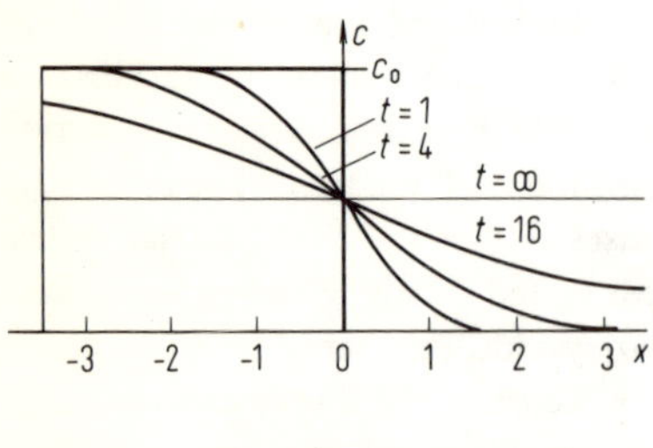

Fig. 5-5. Graphical representation of the solution of Fick's second law for initial conditions $c = c_0\ (x < 0)$ and $c = 0\ (x > 0)$. The curves are concentration profiles for various times. The units are chosen arbitrarily.

Again, the boundaries of the system are considered to lie outside of the diffusion zone. The solution is then obtained simply by superimposing the solutions of an infinite number of point sources lying in the region $x < 0$. The notation is simplified by introducing the so-called error function (erf):

$$\operatorname{erf}(x) = \frac{2}{\pi^{1/2}} \int_0^x \exp(-\eta^2)\,\mathrm{d}\eta\,; \quad \operatorname{erf}(-x) = -\operatorname{erf}(x); \quad \operatorname{erf}(\infty) = 1 \tag{5-41}$$

By integrating over eq. (5-40) and observing the initial and boundary conditions as shown in Fig. 5-5, we obtain the following solution of Fick's second law for this diffusion problem:

$$c(x, t) = \frac{c_0}{2}\left[1 - \operatorname{erf}\frac{x}{2\sqrt{\tilde{D}t}}\right] \tag{5-42}$$

The solution for various diffusion times (or for various values of the parameter $\tilde{D}t$) is shown in Fig. 5-5. Values of the error function (erf) can be found in many tabulations [1, 21].

It may often happen that we wish to calculate the diffusion coefficient $\tilde{D}$ from experimental measurements of the total amount of material which has been transported from the region $x < 0$ to the region $x > 0$. For the initial and boundary conditions of Fig. 5-5, the solution $\int_{x=0}^{\infty} c(x, t)\,\mathrm{d}x$ can be obtained either by direct integration or by the following method: At the point $x = 0$ and time t, the concentration gradient is given by differentiation of eq. (5-42) as $\left(\frac{\partial c}{\partial x}\right)_{x=0} = -\frac{c_0}{2}(\pi\tilde{D}t)^{-1/2}$. The corresponding flux is given by Fick's first law as:

$$j(x = 0) = -\tilde{D}\left(\frac{\partial c}{\partial x}\right)_{x=0} = \frac{c_0}{2}\left(\frac{\tilde{D}}{\pi t}\right)^{1/2} \tag{5-43}$$

Therefore, up to time t, an amount of material equal to

$$s = \int_0^t j\,\mathrm{d}t = c_0\left(\frac{\tilde{D}t}{\pi}\right)^{1/2} \tag{5-44}$$

has diffused per unit cross-section into the semi-infinite space $x > 0$. The chemical diffusion coefficient is thus given by:

$$\tilde{D} = \frac{s^2\pi}{c_0^2 t} \tag{5-45}$$

Another common initial condition occurs when the initial concentration c is equal to c_0 in a finite slice of thickness $2h$ between $x = -h$ and $x = +h$. The solution is once again obtained by integration over the point source solutions as follows:

$$c = \frac{c_0}{2}\left[\operatorname{erf}\left(\frac{h + x}{2\sqrt{\tilde{D}t}}\right) + \operatorname{erf}\left(\frac{h - x}{2\sqrt{\tilde{D}t}}\right)\right] \tag{5-46}$$

Concentration profiles are shown in Fig. 5.6.

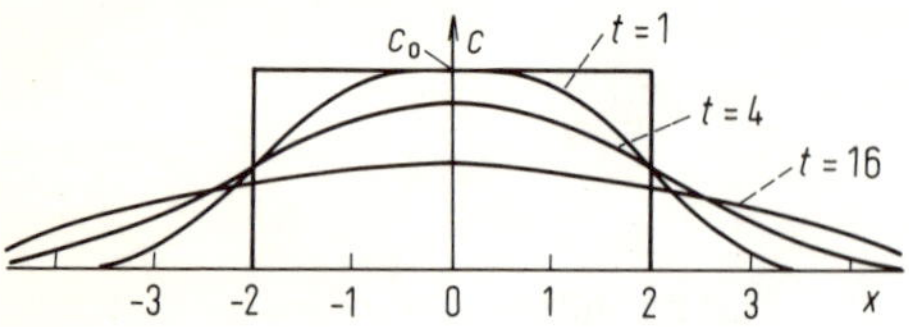

Fig. 5-6. Graphical representation of the solution of Fick's second law with an extended source as initial condition. The units have been chosen arbitrarily. Concentration profiles are given for various times.

To conclude this section, the following problem will be treated because of its practical importance (see Fig. 5-7).

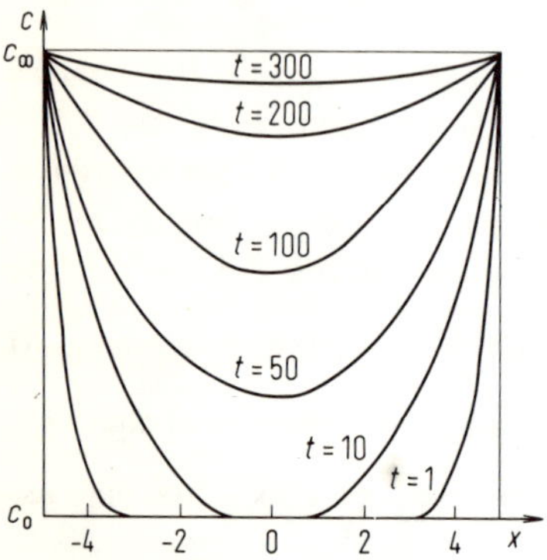

Fig. 5-7. Diffusion into a plate of thickness $2h = 10$ with initial concentration c_0 and a constant surface concentration c_∞. Units of x and t are chosen arbitrarily.

Initially, we have a finite plate with homogeneous concentration c_0. At time $t > 0$ the surface concentration is held constant at c_∞. For finite systems of this type we can no longer integrate over the solutions of point sources, but rather, we obtain solutions in the form of infinite series. In the case of Fig. 5-7, the concentration is given as a function of position and time by the expression:

$$\frac{c - c_0}{c_\infty - c_0} = 1 - \frac{4}{\pi} \sum_{n=0}^{\infty} \frac{(-1)^n}{2n+1} \left[\exp - (2n+1)^2 \frac{\pi^2}{4} \frac{\tilde{D}t}{h^2}\right] \cos\left(\frac{2n+1}{2} \pi \frac{x}{h}\right) \tag{5-47}$$

It is usually advantageous to introduce dimensionless characteristic parameters in order to generalize the mathematical solution. For the present case, the obvious parameters are $T = (\tilde{D}/h^2)\, t$ and $X = x/h$. T and X can be thought of as generalized time and coordinate. From eq. (5-47), then, we have:

$$\frac{c - c_0}{c_\infty - c_0} = 1 - \frac{4}{\pi} \sum_{n=0}^{\infty} \frac{(-1)^n}{2n+1} \left[\exp - (2n+1)^2 \frac{\pi^2}{4} T\right] \cos\left(\frac{2n+1}{2} \pi X\right) \tag{5-48}$$

Integration at constant T for $0 \leq X \leq 1$ gives half of the total amount of material s which has diffused into the plate. s may then be expressed relative to the equilibrium value s_∞ after an infinitely long diffusion time as:

$$s/s_\infty = 1 - \frac{8}{\pi^2} \sum_{n=0}^{\infty} \frac{1}{(2n+1)^2} \left[\exp - (2n+1)^2 \frac{\pi^2}{4} T \right] \tag{5-49}$$

Formulae of this sort are necessary for calculations involving in-gassing or out-gassing processes, carburization or decarburization processes, and decomposition processes in solid lamellar bodies or plates.

5.5.5. Analytical solution of the grain boundary diffusion problem

In section 5.3.5, the process of diffusion along outer surfaces and internal surfaces (grain boundaries) has been discussed. However, the particles of species i, while diffusing along grain boundaries into a solid A, are partly percolating into the bulk of the grains, to an extent which is determined by the bulk diffusion coefficient. In the following section, an approximate solution of this complicated diffusion problem will be discussed. As long ago as 1951, Fisher [28] worked out this approximation in order to explain the measured concentration profiles of radioactive silver which was diffusing from the surface into the interior of polycrystalline silver. The surface activity of radioactive silver was kept constant. The concentration profile was determined by a sectioning technique.

In Fig. 5-8, the diffusion geometry and a three-dimensional concentration profile are shown. It is assumed that the thickness δ of the grain boundary is small compared to the characteristic diffusion length $\sqrt{D_\mathrm{L} t}$ in the bulk of the grains and also to the diffusion length $\sqrt{D_\mathrm{D} t}$ along the grain boundary itself. Furthermore, $D_\mathrm{D}/D_\mathrm{L} \gg 1$. On the basis of the diffusion geometry shown in Fig. 5-8, and because of the above mentioned assumptions, the differential equation for this diffusion problem, where particles i diffuse into the bulk of the material mainly along grain boundaries, reads:

$$\frac{\partial c_i}{\partial t} = D_\mathrm{D} \cdot \frac{\partial^2 c_i}{\partial y^2} + D_\mathrm{L} \left(\frac{\partial c_i}{\partial x} \right)_{x \simeq 0} \cdot \frac{2}{\delta} \tag{5-50}$$

The second term on the right-hand-side of eq. (5-50) takes into account the percolation of particles i into the interior of the grains. As an approximation, it is assumed that this is a one-dimensional process in the x-direction, perpendicular to the grain boundary. As long as $D_\mathrm{D}/D_\mathrm{L} \gg 1$ (i.e. as long as the penetration length in the grain boundary is far greater than the penetration length into the interior of the grain), and as long as the flux j_i into the grain is perpendicular to the grain boundary, the differential equation for diffusion in the grain itself reads simply:

$$\frac{\partial c_i}{\partial t} = D_\mathrm{L} \cdot \frac{\partial^2 c_i}{\partial x^2} \tag{5-51}$$

The result of this percolation into the grain is that, after some time, the concentration distribution in the grain boundary can be approximated as a steady state distribution. This means that $\partial c_i/\partial t$ in eq. (5-50) almost vanishes. If we let $\partial c_i/\partial t$ be zero, then the solution of eq. (5-51) is:

$$c_i(x) = c_i(y) \cdot \left(1 - \operatorname{erf} \frac{x}{2\sqrt{D_\mathrm{L} t}} \right) \tag{5-52}$$

as can be seen from a comparison of Fig. 5-5 and eq. (5-42). We must now explicitly calculate $c_i(y)$ as a function of time. To this end, we introduce eq. (5-52) into eq. (5-50) with the restriction $\partial c_i/\partial t = 0$, and, after some elementary rearrangements, we obtain:

$$\frac{c_i(y)}{c_i(0)} = \exp - \left[\frac{2 \cdot D_L^{1/2}}{\pi^{1/2} \cdot \delta \cdot D_D}\right]^{1/2} \cdot \frac{y}{t^{1/4}} \tag{5-53}$$

It is to be noted that the nature of the approximations excludes the application of eq. (5-53) for times $t \to 0$. As long as the grain size is large compared to the penetration depth into the grain, one obtains the amount s_i of substance i in a penetration depth between y and $y + \mathrm{d}y$ (where y starts at the surface of the grain) by integrating eq. (5-52) between the limits $x = 0$ and $x = \infty$. If one plots s_i (the calculated amount of material i) on a logarithmic scale as a function of penetration depth y, the slope of this curve is given by:

$$\frac{\mathrm{d} \log s_i}{\mathrm{d}y} = -\frac{(2.3) \cdot 2}{\pi^{1/4}} \cdot \frac{D_L^{1/4}}{D_D^{1/2}} \cdot \frac{1}{\delta^{1/2} \cdot t^{1/4}} \tag{5-54}$$

Since the left-hand-side of eq. (5-54) can be determined from experiments, one can derive the grain boundary diffusion coefficient D_D with the help of eq. (5-54), provided that the volume diffusion coefficient D_L is known, and that an assumption has been made concerning the thickness δ of the grain boundary. A characteristic value for δ in metals is $5 \cdot 10^{-8}$ cm.

The above calculations, which are a zeroth order approximation, and which follow essentially the ideas of Fisher [28], can be further refined. Higher order approximations can

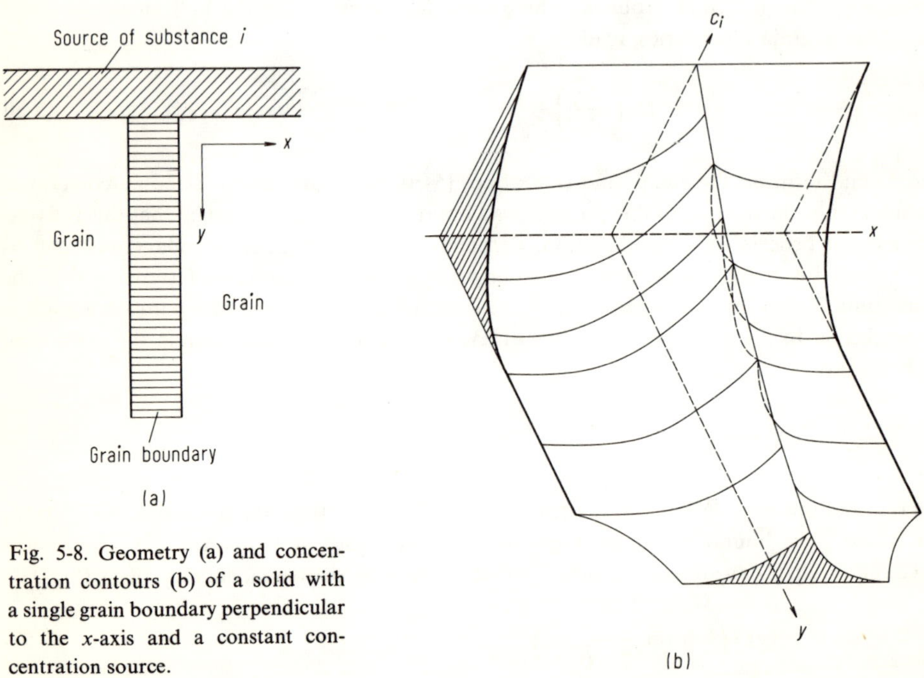

Fig. 5-8. Geometry (a) and concentration contours (b) of a solid with a single grain boundary perpendicular to the x-axis and a constant concentration source.

partly be found in the literature [29-31]. They confirm the validity of the above straight forward solution of the problem inasmuch as the higher order corrections alter the numerical values by a factor of 2 to 3. Despite the fact that the starting point of all calculations are the somewhat unrealistic grain boundary geometries plotted in Fig. 5-8, which do not conform to the grain boundary geometries found in reality, and also despite the approximate character of the concept of a grain boundary thickness δ, the above zeroth order approximation of the grain boundary diffusion problem will suffice in most cases.

5.5.6. Boltzmann-Matano analysis of the concentration profile

In section 5.5.4, Fick's laws were solved for binary systems under the assumption that the chemical diffusion coefficient $\tilde{D}$ is constant. This assumption is never strictly true, and is only approximately true in limiting cases. Thus, we must seek solutions for other cases as well. If we know the concentration dependence of the diffusion coefficient as well as the thermodynamics of the system, we can often reach conclusions regarding the type of disorder and the diffusion mechanism. Therefore, it is especially important in the study of point defect disorder and in the study of the elementary processes of solid state reactions that we know the concentration dependence of the diffusion coefficient.

In the following discussion we shall once again be considering an isothermal, isotropic, one-dimensional diffusion problem. Local thermodynamic equilibrium of defects is assumed. Therefore, $\tilde{D}$ will be a unique function of the composition at every coordinate x.

Boltzmann and Matano showed [22, 23] how the concentration dependent chemical diffusion coefficient $\tilde{D}$ (c) can be determined from the data obtained in a diffusion experiment with two semi-infinite regions and the initial conditions of Fig. 5-5 for the case where the molar volume V_m of the binary system is independent of concentration.

Fick's second law for the case of a concentration dependent chemical diffusion coefficient and the given initial and boundary conditions cannot be explicitly solved even by means of the substitution $y = x/\sqrt{t}$. However, if this substitution is made in eq. (5-31), then the equation

$$\frac{\mathrm{d}}{\mathrm{d}y}\left(\tilde{D}\frac{\mathrm{d}c}{\mathrm{d}y}\right) = -\frac{y}{2}\frac{\mathrm{d}c}{\mathrm{d}y} \tag{5-55}$$

is obtained. In this equation, only total differentials are found. Furthermore, under the initial and boundary conditions of Fig. 5-5 it follows from a single integration that:

$$\tilde{D}\,(c) = -\frac{1}{2}\left(\frac{\mathrm{d}y}{\mathrm{d}c}\right)_c \int_{c_0}^{c} y\,\mathrm{d}c \tag{5-56}$$

Since the diffusion profile was obtained at a fixed time t, we can set $\mathrm{d}y = \mathrm{d}x/\sqrt{t}$, and in place of eq. (5-56) we can write:

$$\tilde{D}\,(c) = \frac{1}{2t}\left(\frac{\mathrm{d}x}{\mathrm{d}c}\right)_c \int_{c}^{c_0} x\,\mathrm{d}c \tag{5-57}$$

The differential quotient $(\mathrm{d}x/\mathrm{d}c)_c$ is taken at the point where the concentration is equal to c. Furthermore, it is evident that all material which has diffused out of the semi-infinite space $x < 0$ must now be found in the semi-infinite space $x > 0$. Therefore, the coordinate of the so-called Matano interface $x_m = 0$ can be calculated by means of the equation:

$$\int_0^{c_0} x \, \mathrm{d}c = 0 \tag{5-58}$$

For the case of constant molar volume, this interface is coincident with the geometrical plane $x = 0$ (in an external coordinate system) which separated the two semi-infinite spaces of the diffusion couple at the beginning of the experiment.

The practical analysis of a diffusion experiment in order to determine the concentration dependent chemical diffusion coefficient $\tilde{D}(c)$ is carried out as follows. First of all, the Matano interface $x_m = 0$ is determined graphically using eq. (5-58). This defines a coordinate system. Then, for every concentration c, the slope $(\mathrm{d}x/\mathrm{d}c)_c$ of the experimental curve $x(c)$ and the area $\int_c^{c_0} x \, \mathrm{d}c$ are graphically determined in this coordinate system. The diffusion coefficient $\tilde{D}(c)$ is then calculated from eq. (5-57). The diffusion time t must be known.

Frequently, in practice, it is not convenient to have to determine the position $x_m = 0$ from eq. (5-58) nor to mark the original concentration discontinuity so that its position can be found again after the diffusion experiment. However, we can avoid the necessity of finding the position $x_m = 0$ by first of all solving equation (5-55) in a general coordinate system $x = x + \alpha$. Then, by introducing the condition of eq. (5-58) into eq. (5-57) and observing the given boundary conditions, it can be shown that the following equation is generally applicable [24]:

$$\tilde{D}(c) = \frac{1}{2t\,(\partial c/\partial x)_x} \left[(1 - c(x)) \int_{-\infty}^{x} c \, \mathrm{d}x + c(x) \int_{x}^{+\infty} (1 - c) \, \mathrm{d}x \right]. \tag{5-59}$$

This formula permits us to calculate the concentration dependent diffusion coefficient $\tilde{D}(c)$ without the necessity of determining the position of the origin of the coordinate system, since the integrals in eq. (5-59) are no longer dependent upon the position of the origin. In part (a) of Fig. 5-9, the calculation of the concentration dependent diffusion coefficient according to eq. (5-59) is illustrated. In Fig. 5-9 part (b), the method of determining the Matano interface is shown.

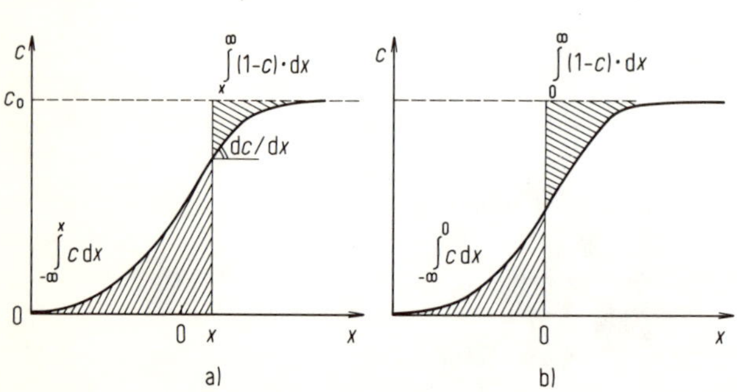

Fig. 5-9. The Boltzmann-Matano analysis of a concentration profile as given in [24].

Different authors have pointed out that, by virtue of its very general assumptions, the Boltzmann-Matano analysis is also applicable to diffusional processes in which more than one phase is initially present in the diffusion zone, or in which additional phases are formed as a result of the diffusional process [19, 25]. In later sections we shall be dealing with solid state reactions in which multiphase diffusional layers are formed. In these sections, the above matter will be examined more closely, and the corresponding equations will be given.

5.6. Literature

General Literature:

[1] W. Jost, Diffusion in Solids, Liquids and Gases, 3rd ed., Academic Press, New York 1960.
[2] J. Crank, Mathematics of Diffusion, Oxford University Press, Glasgow 1956.
[3] H. S. Carslaw and J. C. Jaeger, Conduction of Heat in Solids, 2nd ed., Oxford University Press, Glasgow 1959.
[4] P. G. Shewmon, Diffusion in Solids, McGraw-Hill Book Comp., Inc., New York 1963.
[5] J. R. Manning, Diffusion Kinetics for Atoms in Crystals, D. van Nostrand Comp., Inc., Princeton 1968.

Special Literature:

[6] R. Haase, Thermodynamik der irreversiblen Prozesse, Dr. Dietrich Steinkopff Verlag, Darmstadt 1963, p. 276.
[7] G. E. Rhead, Acta Met. *13*, 223 (1965).
[8] R. A. Oriani and C. A. Johnson in J. O'M. Bockris and B. E. Conway, Modern Aspects of Electrochemistry V, Butterworth Scientific Publ., London 1969, p. 90.
[9] J. Choi and P. G. Shewmon, Trans. AIME *224*, 589 (1962).
[10] R. E. Hoffman and D. Turnbull, J. Appl. Phys. *22*, 634 (1951).
[11] H. J. Queisser, Disc. Faraday Soc. *38*, 305 (1964).
[12] D. Turnbull and R. E. Hoffman, Acta Met. *2*, 419 (1954).
[13] B. S. Chandrasekhar, Rev. Modern Phys. *15*, 1 (1943).
[14] J. Bardeen and C. Herring in W. Shockley, Imperfections in Nearly Perfect Crystals, John Wiley and Sons, Inc., New York 1952, p. 261.
[15] A. D. LeClaire and A. B. Lidiard, Phil. Mag. *1*, 518 (1956).
[16] K. Compaan and Y. Haven, Trans. Faraday Soc. *52*, 786 (1956).
[17] H. Rickert and C. Wagner, Z. Elektrochem., Ber. Bunsenges. phys. Chem. *66*, 502 (1962).
[18] J. R. Wynnyckyj, H. Schmalzried and H.-G. Sockel, Z. phys. Chem. NF *68*, 30 (1969); H.-G. Sockel, H. Schmalzried and J. R. Wynnyckyj, Z. phys. Chem. NF *68*, 49 (1969).
[19] C. Wagner, Acta Met. *17*, 99 (1969).
[20] L. S. Darken, Am. Inst. Mining Met. Engrs. Inst. Met. Div. Metals Technol. *15*, Techn. Publ. 2311 (1948), 2443 (1948).
[21] Jahnke, Emde and Lösch, Tafeln höherer Funktionen, 7th ed., B. G. Teubner Verlagsgesellschaft, Stuttgart 1966.
[22] L. Boltzmann, Wiedemanns Ann. Physik *53*, 959 (1894).
[23] C. Matano, Jap. J. Physics *8*, 109 (1933).
[24] F. Sauer and V. Freise, Z. Elektrochem., Ber. Bunsenges. phys. Chem. *66*, 353 (1962).
[25] Th. Heumann, Z. Metallk. *58*, 168 (1967).
[26] R. H. Condit in T. J. Gray and V. D. Fréchette: Materials Science Research IV, Plenum Press, New York 1969, p. 284.
[27] R. P. Smith, Acta Met. *1*, 578 (1953).
[28] J. C. Fisher, J. appl. Phys. *22*, 74 (1951).
[29] R. T. P. Whipple, Phil. Mag. *45*, 1225 (1954).
[30] H. S. Levine and C. J. McCallum, J. appl. Phys. *31*, 595 (1960).
[31] T. Suzuoka, J. Phys. Soc. Jap. *19*, 839 (1964).

[32] H. Eyring, J. Chem. Phys. *3*, 107 (1935); Chem. Rev. *17*, 65 (1935).
[33] S. Glasstone, K. J. Laidler and H. Eyring, The Theory of Rate Processes, McGraw-Hill Book Comp., Inc., New York 1941.
[34] W. Jost and K. Hauffe, Diffusion, 2nd ed., Dr. Dietrich Steinkopff Verlag, Darmstadt 1972.
[35] S. A. Rice, Phys. Rev. *112*, 804 (1958).

6. Reactions in the solid state – ionic crystals

6.1. Reactions in single-phase systems

Solid state reactions may be sub-divided into heterogeneous reactions and reactions in a single-phase system. The latter can be further sub-divided into homogeneous reactions and reactions which occur in a phase where there are local variations in composition. In the following, reactions of this last type are simply called reactions in an inhomogeneous phase. A homogeneous reaction in a crystal can only occur through a local rearrangement of structural elements, whereas in inhomogeneous systems, local fluxes of the structural elements occur. For example, the equilibration of defects, which is so important in solid state chemistry, is a homogeneous reaction as long as it does not take place on inner or outer crystal surfaces. Quantitative descriptions of diffusional processes are generally based on the assumption of local defect equilibrium. Therefore, an understanding of the kinetics of the processes by which the defect equilibria are maintained is essential to a discussion of diffusion-controlled solid state reactions. An example of a homogeneous reaction in a solid is the equilibration of Frenkel defects following a change in temperature or pressure. An example of a chemical reaction in an inhomogeneous ionic crystal is the readjustment of the concentrations of vacancies and defect electrons in transition metal oxides following a sudden shift in the external oxygen partial pressure. The familiar chemical diffusional processes also belong to this large class of single-phase reactions. In this chapter we shall be concerned mainly with interdiffusional processes in quasi-binary systems in which partial or complete miscibility of the reactant partners occurs. In the following discussions, systems such as KCl-RbCl, AgBr-NaBr, or MgO-NiO will be treated in detail.

6.1.1. Homogeneous reactions

In the discussion of defect thermodynamics in chapter 4, it was shown that the equilibrium constants for defect equilibrium are dependent upon both temperature and pressure. Accordingly, if one of these intensive thermodynamic variables is changed for a homogeneous ionic crystal which was originally in thermodynamic equilibrium, then the defect concentrations will attempt to reach a new equilibrium value as rapidly as is kinetically possible. Depending upon the type of disorder, the equilibration occurs by means of reactions: 1. between the defects themselves, 2. between defects and lattice atoms at sites of repeatable growth (inner and outer interfaces, dislocations), or 3. between defects and the components of the crystal on the outer crystal surface. Strictly speaking, only the first case can be called a homogeneous reaction. The relaxation time τ for defect equilibration is defined as the time required for the defects to approach their new equilibrium concentration within a fraction $1/e$. The relaxation time depends both upon the mobility of the defects and upon the density and geometrical arrangement of the defect sources and sinks. Even if the defects have arrived at the reaction site by diffusion, it is nevertheless quite possible that an activation energy is still required for the final reaction step between defects or between defects and lattice atoms at sites of repeatable growth. The same is true if defects are created at sites of repeatable growth before they diffuse into the bulk volume.

There are three distinguishable cases of homogeneous reactions between point defects: 1. Two different point defects come together either because of their random diffusional motion or because in addition an attractive force is acting. Upon coming together, they combine to form a normal lattice particle. The best known example of this is the reaction between the two Frenkel defects in silver bromide. The overall reaction equation was given in eq. (4-19): $Ag_i^{\bullet} + V'_{Ag} = Ag^x_{Ag} + V^x_i$. 2. Two similar or dissimilar defects come together to form a defect associate. For example, in alkali halides, reactions can occur between dissolved alkaline earth ions and vacancies according to the reaction equation: $Ca^{\bullet}_K + V'_K = [CaV]^x$. Here, a neutral complex $[CaV]^x$ is formed. 3. Ions in different sublattices exchange positions. The reaction for this case is: $A_A + B_B = A_B + B_A$. This class of homogeneous reactions also includes exchanges of charge between ions, as for instance: $A^{n+}_A + B^{m+}_B = A^{(n-1)+}_A + B^{(m+1)+}_B$. The last two reactions mentioned above play an important role in the preparation of oxide ferrites having closely controlled electrical and magnetic properties [1].

It is frequently of practical value to be able to approximately calculate the relaxation time τ for the equilibration of defects. Ionic crystals are generally prepared at temperatures so high that equilibrium among defects occurs almost instantaneously. However, the crystal is later used mostly at room temperature. During cooling from the preparation temperature to room temperature, the defect equilibria will be frozen in at some point, and the concentration of defects at room temperature will therefore be fixed at some high temperature equilibrium value. The optical, electrical, magnetic, and reactive or catalytic properties of crystals are influenced or determined by the point defects. Therefore, an understanding of defect relaxation is essential for proper temperature programming during the cooling of a crystal in order to obtain the desired properties.

Defect relaxation times for homogeneous reactions in solids can be calculated essentially by the methods of homogeneous chemical kinetics [2, 3]. For the sake of illustration, let us consider more closely the equilibration of Frenkel defects in silver bromide following a sudden change in temperature.

According to eq. (4-19), the time derivative of the interstitial and vacancy concentrations is assumed to be given by the expression:

$$\frac{\mathrm{d}c_{Ag_i^{\bullet}}}{\mathrm{d}t} = \frac{\mathrm{d}c_{V'_{Ag}}}{\mathrm{d}t} = -k_1 c_{Ag_i^{\bullet}} c_{V'_{Ag}} + k_2 c_{Ag^x_{Ag}} \tag{6-1}$$

The variables c_i are molar concentrations per cm^3. Eq. (6-1) says that Frenkel pairs are annihilated at a rate proportional to the frequency with which vacancies and interstitial ions come together, while the rate of formation of Frenkel pairs is proportional to the concentration of lattice ions Ag^x_{Ag}. The concentrations of defects are very small, so that the number of lattice ions Ag^x_{Ag} remains constant during the reaction. Furthermore, since the condition $c_{Ag_i^{\bullet}} = c_{V'_{Ag}}$ holds for the case of Frenkel type disorder, eq. (6-1) may be rearranged to read:

$$\frac{\mathrm{d}c_{Ag_i^{\bullet}}}{\mathrm{d}t} = -k_1 (c^2_{Ag_i^{\bullet}} - K); \quad K = c^2_{Ag_i^{\bullet}}\,(\mathrm{eq}) \tag{6-2}$$

K is the equilibrium constant for Frenkel disorder. This can be seen if we set the time derivative in eq. (6-2) equal to zero, since this is the condition for thermodynamic equilibrium. On the basis of the simple theory of homogeneous reactions, two limiting cases arise. 1. The electrostatic interaction between electrically charged defects at a separation r_{1-2} is negligible compared to the thermal energy kT. That is: $\frac{e_1 e_2}{\varepsilon \varepsilon_0 r_{1-2}} \ll kT$ where e_1 and e_2 are the electric charges of the defects, and ε and ε_0 are the relative and absolute dielectric constants respectively. We then obtain the following expression for k_1 :

$$k_1 = 4\pi a \,(D_{\mathrm{Ag_i^{\bullet}}} + D_{\mathrm{V'_{Ag}}})\, N_0 \tag{6-3}$$

where a is the separation of the defects just before the final step leading to their mutual annihilation, assuming that an activation is no longer necessary at this point. N_0 is Loschmidt's number. The limiting case of negligible electrostatic interaction which is quantitatively formulated in eq. (6-3) is only applicable for very small defect concentrations ($N < 10^{-6}$). 2. The electrostatic energy $\frac{e_1 e_2}{\varepsilon\, \varepsilon_0 r_{1-2}}$ is no longer negligible compared to $k\,T$. In this case, k_1 is given by [8]:

$$k_1 = 4\pi \,(D_{\mathrm{Ag_i^{\bullet}}} + D_{\mathrm{V'_{Ag}}})\, N_0 \,\frac{e_1 e_2}{\varepsilon\, \varepsilon_0 k\, T} \tag{6-4}$$

By integrating eq. (6-2) we obtain an expression for the instantaneous deviation from defect equilibrium. After a sufficient time, this becomes proportional to exp $(-t/\tau)$. The relaxation time τ is then given from eqs. (6-2) to (6-4) as:

$$\tau = \frac{1}{2k_1 K^{1/2}} \tag{6-5}$$

By introducing numerical values for silver bromide into eq. (6-5), we find that even at room temperature the relaxation time for the equilibration of Frenkel defects is of the order of milliseconds and less. Because of the paucity of available data [9], only such rough estimates can be made. Furthermore, it should be remembered that these estimates are dependent upon the assumption that the defect reaction is, in fact, a homogeneous reaction.

At this point let us briefly consider the formation of associates. The formation of associates between cation vacancies and divalent impurities in alkali halides has already been given as an example. Such reactions are homogeneous solid state reactions, and so the relaxation time for the formation of associates can be calculated in a completely analogous manner to the calculation of the relaxation time for the equilibration of Frenkel defects. The result of such calculations is precisely the same as the result given in eq. (6-5). It is only necessary, in the case of association, to replace the concentration c (eq) $= K^{1/2}$ in the denominator by the nearly constant concentration of the corresponding majority defect. In general, in the case of the formation of defect associates, we can conclude that the equilibrium concentration is attained rapidly compared to the time required by defect reactions which occur at sites of repeatable growth.

Finally, let us discuss in somewhat more detail the rearrangement of cations between the various sublattices of a crystal when the equilibrium is disturbed. As an example, let us consider the reaction $\mathrm{Ni_A^{2+}} + \mathrm{Al_B^{3+}} = \mathrm{Ni_B^{2+}} + \mathrm{Al_A^{3+}}$ occurring in nickel aluminate, $\mathrm{NiAl_2O_4}$, a phase which has the spinel structure. The subscripts A and B denote sites of tetrahedral and octahedral symmetry within the close-packed face-centred-cubic framework formed by the oxygen ions. This reaction would take place following a disturbance of the distribution equilibrium among the cations in the sublattices. Such a disturbance might be caused by a sudden change in temperature. The equilibrium constant $K(T) = \mathrm{e}^{-\Delta G^0/RT}$ of the exchange reaction is temperature dependent. ΔG^0 is the standard free energy of the reaction. Measurements of the concentrations of Ni_A^{2+} and Ni_B^{2+} using optical absorption techniques have shown that ΔG^0 is about -2.6 kcal. If the temperature of a crystal which is in equilibrium is suddenly changed, then the cation distribution will gradually approach a new equilibrium value. The exchange of ions between the two sublattices is analyzed as follows: In order that a Ni_A^{2+} ion can take up a position on an empty site in the B-sublattice, it must first leave the A-sublattice and diffuse through the interstices until it becomes a nearest neighbour of a vacancy V_B. There is then a certain probability that it will jump into this vacancy. The exchange of the particles Al_B^{3+}, Ni_B^{2+}, and Al_A^{3+} can be similarly described. The overall exchange is thus a complicated coupled process. The partial steps of this process are, of course, homogeneous reactions, and can be formulated as in eq. (6-1). Although the overall reaction is a very simple substitution mechanism, the general mathematical treatment of the kinetics is very difficult. Only for the case of very long relaxation times can the reaction be described by a single relaxation process $e^{-t/\tau}$. This then can be taken to be the relaxation time of the slowest partial step in the coupled reaction.

6.1.2. Reactions in inhomogeneous media

The quantitative description of most diffusional processes in single-phase media is based upon the assumption of local defect equilibrium. Therefore, in this section we shall first of all examine processes by which defect equilibrium is attained. In this regard, two important fundamental types of defect reactions should be discussed: 1. Following a sudden change in temperature or pressure, a reaction between defects and lattice atoms occurs at sites of repeatable growth such as inner or outer surfaces and dislocations. For a crystal with Schottky disorder, the reaction equation may be formulated as:

$$V'_{Me} + V^{\bullet}_{X} + MeX_{defect} = Me^{x}_{Me} + X^{x}_{X} \tag{6-6}$$

In words, eq. (6-6) says that an excess of vacancies of the Schottky disorder type react with lattice molecules situated at crystal defects (sites of repeatable growth) to form regular ions in the crystal. If local thermodynamic equilibrium is continually maintained at the sites of repeatable growth during this process, then we can make calculations regarding the formation and annihilation of point defects, provided that we know the number and spatial arrangement of these sites. In principle, this problem has already been solved in section 5.5.4. 2. The partial pressure of X_2 in contact with a crystal MeX which was initially in equilibrium is suddenly changed. The new equilibrium defect concentrations, which are dependent upon the activities, are first attained on the surface. As a result, a concentration gradient of defects is set up from the surface to the interior. This concentration gradient becomes evened out through diffusion. This problem can also be solved for a known geometry and for known defect mobilities by the methods of chapter 5. By way of illustration, let us work through one example for each case.

(1) If the temperature of a potassium chloride crystal is suddenly decreased, then the number of Schottky defects present will be greater than the new equilibrium value. That is, the crystal is supersaturated with respect to defects. Equilibration is then achieved by the reaction:

$$V'_{K} + V^{\bullet}_{Cl} + KCl_{defect} = K^{x}_{K} + Cl^{x}_{Cl} \quad \text{(in KCl)} \tag{6-6a}$$

Let us assume that the heterogeneous reaction (6-6a) takes place on infinitely extended laminar plates. In the zeroth approximation, this can be taken as a description of the mosaic structure of the crystal. The fluxes of vacancies in the cation and anion sublattices are given by:

$$j_- = -\frac{c_{V_K'} D_{V_K'}}{RT} \cdot \text{grad}\, \eta_{V_K'}\,; \quad j_+ = -\frac{c_{V_{Cl}^{\bullet}} D_{V_{Cl}^{\bullet}}}{RT} \cdot \text{grad}\, \eta_{V_{Cl}^{\bullet}} \tag{6-7}$$

Because of the condition of electroneutrality, the local concentrations $c_{V_K'}$ and $c_{V_{Cl}^{\bullet}}$ must be equal to each other at all times, even during the equilibration reaction. Therefore, j_+ is always equal to j_-. By using this condition, we can eliminate the diffusion potential ϕ from the flux equations. Furthermore, the laws of ideal dilute solutions may be used in the present case. Setting $\eta_i = \mu_i + z_i F\phi$ and $\mu_i = \mu_i^0 + RT \ln c_i$, we obtain the following equation for the flux:

$$j = j_+ = j_- = -\frac{2 D_{V_K'} D_{V_{Cl}^{\bullet}}}{D_{V_K'} + D_{V_{Cl}^{\bullet}}} \cdot \frac{\mathrm{d} c_{V_K'}}{\mathrm{d}x} \tag{6-8}$$

$\dfrac{2 D_{V_K'} D_{V_{Cl}^{\bullet}}}{D_{V_K'} + D_{V_{Cl}^{\bullet}}}$ is the chemical diffusion coefficient $\tilde{D}$. Fick's second law can then be written as:

$$\frac{\partial c_{V_K'}}{\partial t} = \frac{2 D_{V_K'} D_{V_{Cl}^{\bullet}}}{D_{V_K'} + D_{V_{Cl}^{\bullet}}} \cdot \frac{\partial^2 c_{V_K'}}{\partial x^2}$$

For the assumed geometry of the system, the initial and boundary conditions for lamellar plates of thickness $2a$ are:

$$c = c\,(0); \quad -a \leq x \leq +a; \quad t = 0$$
$$c = c\,(\text{eq}); \quad x = \pm a; \quad t > 0$$

The solution of this equation has already been given in eq. (5-47). For times greater than $3a^2/2\pi^2 \tilde{D}$, the series can be terminated after the first term without introducing serious errors. The relaxation time is then given as:

$$\tau = \frac{4a^2}{\pi^2 D_{V_K'} D_{V_{Cl}^{\bullet}}} (D_{V_K'} + D_{V_{Cl}^{\bullet}}) \tag{6-9}$$

τ is proportional to the square of the separation $2a$ of the mosaic grain boundaries. If one of the two defect diffusion coefficients is much greater than the other, then only the smaller of the two remains in the denominator. The relaxation time is then inversely proportional to the diffusion coefficient of the slower defect.

As well as low-angle grain boundaries, dislocation lines also serve as important sources and sinks for defects. The following parameters are convenient for describing the spatial configuration of dislocations: α = radius of the core of a dislocation, β = radius of influence of a dislocation upon defects. For a disordered dislocation structure it is only possible to approximate β as follows:

$$\beta^2 \pi = \frac{1}{\varrho_D}; \quad \beta = \left(\frac{1}{\varrho_D \pi}\right)^{1/2} \tag{6-10}$$

where ϱ_D is the experimentally measurable dislocation density (number of dislocations per cm^2) [10]. The differential equation for the problem can be derived from the flux equation (6-8). In cylindrical coordinates:

$$\frac{\partial c}{\partial t} = \frac{2 D_{V_k'} D_{V_{Cl}^{\bullet}}}{D_{V_k'} + D_{V_{Cl}^{\bullet}}} \cdot \frac{1}{r} \frac{\partial}{\partial r}\left(r \frac{\partial c}{\partial r}\right) \tag{6-11}$$

with initial and boundary conditions:

$$c = c\,(0); \quad \alpha < r < \beta; \quad t = 0$$

$$c = c\,(\text{eq}) \text{ for } r = \alpha; \quad \frac{\partial c}{\partial r} = 0 \text{ for } r = \beta; \quad t > 0$$

where c is the defect concentration. The detailed solution of this partial differential equation can be found in the literature [29]. With a value $\beta/\alpha = 10^3$, which is representative of that found in practice, and for times which are not too small, the relaxation time is given as:

$$\tau \simeq \frac{\gamma}{\pi \varrho_D D_{V_k'} D_{V_{Cl}^{\bullet}}} (D_{V_k'} + D_{V_{Cl}^{\bullet}}) \tag{6-12}$$

where γ is a numerical factor of the order of unity. This equation may be compared with eq. (6-9) which gives τ for the case where the defect annihilation reaction (6-6a) occurs on parallel grain boundaries with separation $2a$.

Two final remarks should be made concerning the validity of this approximate calculation. (1) If point defects are created or annihilated at a dislocation line, this line moves through the lattice (climb). It is assumed that the distance of this nonconservative movement is small compared with β. (2) Because of the cylindrical symmetry of the problem, the main diffusion resistance (i.e. the highest concentration gradient of the defects) is found in the immediate neighbourhood of the dislocation line. Therefore the specific spatial distribution of the dislocation lines is not very important for the given estimate of the relaxation time τ.

(2) Suppose that a thin monocrystalline plate of nickel oxide is initially in equilibrium at 1 000 °C with the oxygen in the air, and that the partial pressure of oxygen is suddenly lowered by a factor of ten. The following defect reaction equation can be written:

$$\tfrac{1}{2}\,O_2 + 3\,Ni_{Ni}^{2+} = 2\,Ni_{Ni}^{\bullet} + V_{Ni}'' + NiO \tag{6-13}$$

For the case in which equilibrium is maintained at the surface of the plate, both the equilibrium vacancy concentration and the concentration of electron holes are instantaneously decreased by a factor of about 0.7 according to the equilibrium condition: $(Ni_{Ni}^{\bullet}) = 2\,(V_{Ni}'') \propto p_{O_2}^{1/6}$. There will be a coupled flux of vacancies and electron holes from the interior to the surface. The equations for these fluxes are:

$$j_+ = -\frac{c_{h^\bullet} D_{h^\bullet}}{RT} \cdot \text{grad}\, \eta_{h^\bullet}; \quad j_- = -\frac{c_{V_{Ni}''} D_{V_{Ni}''}}{RT} \cdot \text{grad}\, \eta_{V_{Ni}''} \tag{6-14}$$

where $h^{\bullet}$ is an electron hole (which can also be written as $Ni^{\bullet}_{Ni}$ or as Ni^{3+}_{Ni}). If the new equilibrium of the crystal with the surrounding oxygen atmosphere is maintained at the surface during the entire diffusional process, then the diffusion problem is completely analogous to the aforementioned case of the annihilation of Schottky defects on parallel planar mosaic grain boundaries. Once again, the electrical potential may be eliminated by means of the coupling condition $j_+ = 2j_-$ which arises because of the necessity of maintaining local electroneutrality. The flux equation is then:

$$j_+ = -\frac{3\,D_{h^{\bullet}} D_{V''_{Ni}}}{D_{h^{\bullet}} + 2\,D_{V''_{Ni}}} \cdot \frac{dc_{h^{\bullet}}}{dx} \tag{6-15}$$

The solution of this problem is exactly the same as the solution given in eq. (6-9) for the equilibration of Schottky defects at mosaic grain boundaries. It is only necessary to set the chemical diffusion coefficient $\tilde{D} = \dfrac{3\,D_{h^{\bullet}} \cdot D_{V''_{Ni}}}{D_{h^{\bullet}} + 2\,D_{V''_{Ni}}}$. Finally, since nickel oxide is a semiconductor: $D_{h^{\bullet}} \gg D_{V''_{Ni}}$, and so $\tilde{D} = 3\,D_{V''_{Ni}}$. The chemical diffusion coefficient is thus equal to three times the diffusion coefficient of the cation vacancies. The slow diffusion of these vacancies determines the rate at which defect equilibrium is approached. The factor 3 arises because the electron holes, by virtue of their high mobility, advance faster than the vacancies, thereby setting up a diffusion potential and accelerating the diffusion of the vacancies in the electrical field.

One final important example of a defect relaxation process should be mentioned in this section. This is the recombination of excess electrons and electron holes in elemental and compound semiconductors following optical excitation (i.e. absorption of light). Very small concentrations of impurities can greatly reduce the relaxation time for this process, since the dissolved impurity atoms (e.g. Ni in Ge) can act as recombination centers.

6.1.3. Reactions in quasi-binary systems

If we were to permit a KCl crystal to react with a RbCl crystal, or a NiO crystal with a MgO or CoO crystal, or a $CoAl_2O_4$ crystal with a $MgAl_2O_4$ crystal, then different situations would be observed in each case, depending upon the relative mobilities of the ionic and electronic defects. First of all, since these systems form nearly ideal solid solutions [34], the thermodynamic factor $(1 + d\ln \gamma_i / d\ln c_i)$ in eq. (5-28) can be set equal to unity, and Fick's first law applies in its simplest form (see eq. (5-29)). Essentially, the problem in this class of reactions is to determine the chemical diffusion coefficient as a function of composition.

The chemical diffusion coefficient can be experimentally determined as described in section 5.5.6 with the aid of the Boltzmann-Matano analysis. The atomistic interpretation of this diffusion coefficient will be illustrated by two examples.

(1) Interdiffusion in the system NiO-MgO.

A strongly asymmetric diffusion profile has been observed after a certain diffusion time by means of electron probe microanalysis. This profile is shown in Fig. 6-1 [11].

The chemical diffusion coefficient can be calculated from this diffusion profile with the aid of the Boltzmann-Matano analysis. It has the form $\tilde{D} = \tilde{D}(N_{MgO} = 0) \cdot \exp - b N_{MgO}$, where $\tilde{D}(N_{MgO} = 0)$ and b are constants. This experimentally determined diffusion coefficient may be interpreted as follows. The fluxes of nickel ions, magnesium ions, and electron holes relative

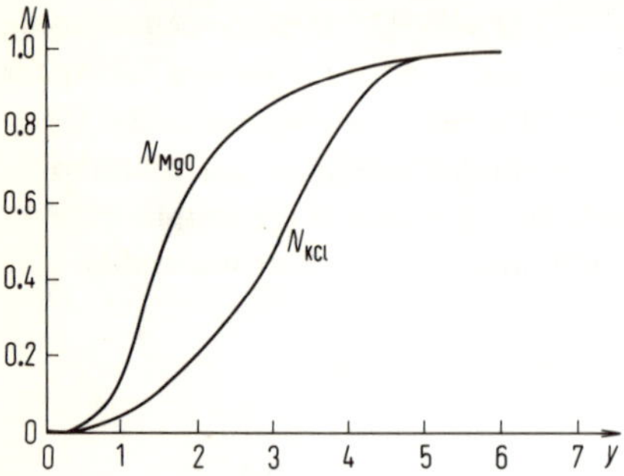

Fig. 6-1. The diffusion profiles in the diffusion couples NiO-MgO ($T = 1370$ °C) and RbCl-KCl ($T = 660$ °C) given in arbitrary units of $y = x/t^{1/2}$. N = mole fraction.

to the oxygen ion sublattice can be described by means of flux equations of the form $j_i = -(c_i D_i/RT) \cdot \text{grad}\,(\mu_i + z_i F\phi)$, where $i = Mg^{2+}$, Ni^{2+}, $h^{\bullet}$. The large oxygen ions are nearly immobile compared to the cations. If the electron hole conduction in the semiconductor solid solution is much greater than the ionic conduction, then it will not be possible for a diffusion potential to be built up as a result of the interdiffusion of cations. The chemical diffusion coefficient can then be calculated from the flux equations in a manner analogous to that described in section 5.5.3. An equation corresponding to the Darken formula (5-37) results:

$$\tilde{D} = D_{\text{Mg}}\,(1 - N_{\text{MgO}}) + D_{\text{Ni}}\,N_{\text{MgO}} \tag{6-16}$$

As always, electroneutrality of the volume elements and local thermodynamic equilibrium have been assumed. In the analogous systems MgO-CoO and NiO-CoO, it is known [12] that, at constant oxygen partial pressure, the mole fraction of vacancies depends upon the mole fraction of cobalt oxide, essentially according to:

$$N_{\text{V}} = N_{\text{h}^{\bullet}}/2 = N_{\text{V}}^{\circ} \exp\left(-\,b\,(1 - N_{\text{CoO}})\right) \tag{6-17}$$

Thus, from eqs. (6-16) and (6-17) it is possible to theoretically interpret the experimentally determined chemical diffusion coefficient if (a) we assume that diffusion of Mg ions and Ni ions occurs via vacancies in the cation sublattice, (b) there is an exponential dependence of the vacancy concentration upon the mole fraction of magnesium oxide in the solid solution NiO-MgO at constant partial pressure of oxygen, and (c) the jump frequencies of the two cations are not very different from one another. Only with these assumptions is it possible to simplify eq. (6-16) to give the expression $\tilde{D} = \tilde{D}\,(N_{\text{MgO}} = 0) \cdot \exp(-\,b\,N_{\text{MgO}})$ which has been found experimentally. These conclusions are supported by the following observations: Firstly, in other oxide solid solutions such as Cr_2O_3-Al_2O_3 or $CoAl_2O_4$-$MgAl_2O_4$, an exponential dependence of the chemical diffusion coefficient (and of the vacancy concentration) upon composition has also been found [35]. Secondly, the tracer diffusion coefficients of Ni and Mg in NiO are almost equal.

It should be noted that the assumption of local defect equilibrium may not hold in the very early stages of the interdiffusion process when steep concentration gradients occur, especially if only the outer crystal surface acts as source or sink for point defects. One would expect then time dependent $\tilde{D}(c)$-values and nonplanar equiconcentration surfaces.

(2) Interdiffusion in the system KCl-RbCl [13].
The solid solution KCl-RbCl differs basically from the solid solution NiO-MgO in two ways. Firstly, the system KCl-RbCl exhibits purely ionic conduction. The transport numbers of electronic charge carriers are negligibly small. Secondly, a finite transport of anions occurs. Because of these facts, the atomic mechanism of the solid state reaction between KCl and RbCl is essentially of a different sort than that between NiO and MgO. Once again, the diffusion profile exhibits an asymmetry (see Fig. 6-1). However, in this case the asymmetry arises not so much because of the variation of the defect concentration with composition, but rather because of the different mobilities of the ions at given concentration. Were the transport number of the chloride ions negligible, then the diffusion potential (which would be set up because of the different diffusion velocities of potassium and rubidium) would ensure that the motion of the two cations is coupled. If, on the contrary, the transference number of the chloride ions is one, then there is no diffusion potential, and the motion of the two cations is decoupled.

If the ionic mobilities (or the component diffusion coefficients) are known, then it is possible to calculate the diffusion profile, as well as the displacement of inert markers, if we assume that local thermodynamic equilibrium is maintained and that the markers are firmly bound to the anion lattice. In order to calculate the chemical diffusion coefficient, we start once again with the flux equations of section 5.2 and eliminate the diffusion potential using the condition of electroneutrality. For an ideal solid solution we obtain the equation [13]:

$$\tilde{D} = (1 - N_{KCl})\, D_K + N_{KCl} D_{Rb} - N_{KCl}(1 - N_{KCl})(D_K - D_{Rb})^2 \cdot \frac{F^2}{RTV\sigma} \tag{6-18}$$

where F is the Faraday constant, V is the molar volume, and σ is the total electrical conductivity. Eq. (6-18) can be rearranged with the help of eq. (5-16) as

$$\tilde{D} = \frac{D_K D_{Rb} + D_{Cl}\,[(1 - N_{KCl})\, D_K + N_{KCl} \cdot D_{Rb}]}{D_{Cl} + N_{KCl} D_K + (1 - N_{KCl}) \cdot D_{Rb}} \tag{6-18a}$$

By letting $D_{Cl} \gg D_K$ and D_{Rb}, eq. (6-18a) yields

$$\tilde{D} = (1 - N_{KCl})\, D_K + N_{KCl} D_{Rb} \tag{6-18b}$$

which is a Darken-type equation as given in eq. (6-16) or (5-37). By letting $D_{Cl} \ll D_K$ and D_{Rb}, eq. (6-18a) yields

$$\tilde{D} = \frac{D_K D_{Rb}}{N_{KCl} D_K + (1 - N_{KCl}) \cdot D_{Rb}} \tag{6-18c}$$

which is the well known Nernst-Planck expression for binary chemical diffusion of charged particles.

If we compare the explicit expression for the chemical diffusion coefficients in eqs. (6-16) and (6-18), we can see how great an influence the mobilities of the structural elements can have on the form of these equations, even when the lattice structure and the diffusion mechanism are the same.

A displacement of inert markers (see page 111) relative to their initial position can occur as a result of the local production and annihilation of defects during a solid state reaction. This phenomenon is called the "Kirkendall effect". This effect will be discussed later in greater detail in connection with reactions between metals. In complete analogy to the case of metals, it is to be expected that a Kirkendall effect will also be observed during interdiffusion in ionic crystals if this is understood to mean a local formation or annihilation of lattice defects which occurs to maintain local defect equilibrium when the ionic mobilities are different. However, in contrast to the case of metallic phases, this need not lead to a movement of inert markers, since in certain cases the inert markers may be firmly anchored to an immobile sublattice. The term "Kirkendall effect" must therefore be applied with care. This term should not simply be used to mean the displacement of inert markers in a reference system which is fixed outside the diffusion zone. For binary metallic systems, such a displacement could occur merely because of a concentration dependence of the molar volume, and for an ionic crystal, as has just been mentioned, such a displacement might not occur at all, even though local defect formation or annihilation is taking place in a sublattice during the interdiffusion.

The last two examples show that interdiffusion processes and chemical diffusion coefficients can vary widely, depending upon the transport numbers of the ionic and electronic defects. A theoretical calculation is only possible if it is assumed that defect equilibrium is maintained. Whether the assumption of local defect equilibrium is applicable to an individual case will depend upon the relaxation time for the defect equilibration process. That is, it will depend upon the density of defect sources and sinks. In most cases, therefore, it will depend upon the density of dislocations and of low- and high-angle grain boundaries.

Reactions in quasi-binary systems in which the reactant partners no longer form a complete range of solid solutions can be treated in single-phase regions in exactly the same way as the diffusional processes just discussed above. This will be discussed in greater detail in connection with reactions in multiphase systems.

Calculations involving diffusion processes in inhomogeneous multicomponent ionic systems have been recently performed by Kirkaldy [30] and Cooper [38]. They worked with the same assumptions that have been made in this section in which quasi-binary systems have been discussed: constant molar volume of the solid solution, and independent fluxes of ions, which are coupled only by the electrical diffusion potential. The latter can be eliminated by the condition $\sum z_i j_i = 0$ which means that local electroneutrality prevails. With these assumptions, and with a knowledge of the thermodynamics of the multicomponent system (which is a knowledge of the activity of the electroneutral components as a function of composition), the individual ionic fluxes can be calculated explicitly with the help of the ionic mobilities and the activity coefficients of the components.

The problem of ionic diffusion in multicomponent systems can also be treated formally with the help of eqs. (5-2). The phenomenological conditions (5-13) and the aforementioned restrictive assumptions allow the calculation of generalized diffusion coefficients D_{ik}. This will be explained fully in section 7.1.3 where diffusion in metallic multicomponent systems is treated. Cooper [38] has successfully used the rules of the matrix calculus in order to transform the general transport coefficients L_{ik} into the generalized diffusion coefficients D_{ik}. In this way, a simple handling of the complicated algebra has become possible. An explanation of the matrix calculus is beyond the scope of this book. The reader interested in this elegant method should consult the original literature [38].

6.2. Reactions in multiphase systems

The simplest reaction of this class occurs when two solid phases A and B form a solid reaction product C. A and B can be either elements or compounds. The essential point is that the reactants A and B are spatially separated from each other by the solid reaction product C. During the course of the reaction, particles will be transported over phase boundaries and through the phases. Thus, there are several different reaction steps to be discussed. The overall driving force for the reaction is the difference in free energy between the reactants and the reaction product. In this section we shall be considering only reactions between crystalline solids, and so the entropies of reaction will be small. Therefore, heat will be liberated during the reaction. For reactions between single crystals where there is a relatively small reaction interface (phase boundary), the heat production per unit time at the reaction front will be very small because of the slow reaction rates. Therefore, the assumption of an isothermal reaction can be made. However, the situation can be quite different in the case of powder reactions where the reaction interface is large. Here the exothermic reaction can result in a self-heating of the reacting particles. This in turn causes an increase in the reaction rate which then leads to a further acceleration in the production of heat. Either a stable or an unstable thermal state can arise, depending upon the rates of production and removal of heat. Heat can be conducted away partially into the interior of those particles which have not yet taken part in the reaction, and partially through the reaction layer and then into the surrounding gas atmosphere by means of radiation and convection. It would be hopeless to try to quantitatively treat this problem for a powder reaction with the most general boundary conditions. However, a model calculation has recently appeared in the literature [31] for the roasting of a single spherical particle of ZnS according to the reaction equation: $ZnS + 3/2\,O_2\,(g) = ZnO + SO_2\,(g)$, $\Delta H = -115$ kcal/mole ($T = 1200$ °C), in which non-isothermal effects have been included. This calculation can be easily applied to classical solid state reactions. The reference to the literature will suffice at this point.

For purposes of a quantitative discussion, let us first make the following assumptions: A and B are binary compounds in a ternary system. The phases in the reaction product lie in the Gibbs triangle on the line joining A and B (see Fig. 6-2).

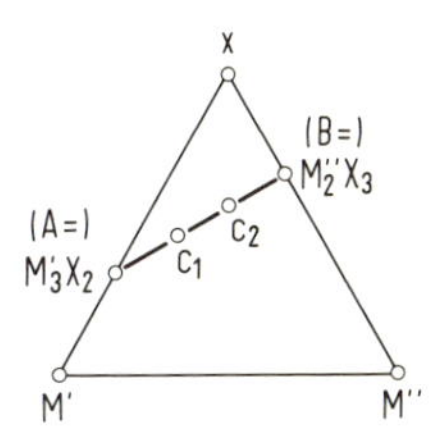

Fig. 6-2. Gibbs triangle for the ternary system M'-M''-X. The quasi-binary line $M_3'X_2$-$M_2''X_3$ is drawn with the product phases C_1 and C_2.

In this way we have limited ourselves to a quasi-binary system. Unless interrupted externally, the reaction will continue until at least one of the two reactants is completely consumed (Gibbs phase rule). If the reaction is interrupted at a time when both reactants are still present, then the reaction product will consist of a sequence of all those phases which occur in the phase diagram as compounds between the reactants A and B in the quasi-binary system, as long as local equilibrium is maintained. It can be seen that we have here a convenient method for investigating quasi-binary lines in ternary phase diagrams.

Suppose, for example, that two binary oxides react to form one or more ternary oxides as in Fig. 6-2. In order to properly understand the reaction, we must first be completely clear as to the number of independent thermodynamic variables. For a given total pressure and a given reaction temperature, there will still be one more independent variable in a binary system, and two more independent variables in a ternary system. Therefore, the experimental conditions are only completely defined if the activity of the component common to all compounds (i.e. the oxygen activity in this case) is fixed in the reaction zone. Only then are the local chemical potentials of the components and the transport coefficients (which depend, in general, upon these potentials) uniquely determined.

6.2.1. Spinel formation reactions

Of all solid state reactions, the formation of oxide spinels is at present the most thoroughly investigated [4, 5, 33]. The first reason for this is the relatively simple crystallographic structure of the spinel lattice. Essentially, this consists of a nearly close-packed face-centred-cubic sublattice of oxygen ions. The tetrahedral and octrahedral interstices of this sublattice are filled in a certain way by the cations. The second reason is that spinels are technically very interesting substances, and one would like to be able to find optimal methods for their preparation. For instance, ferrites are used as control or circuit elements in the electronics industry, and chromite brick is used as cladding in ovens which are used for the production of steel. Therefore, the formation of spinels will now be discussed in detail as a model of a classical solid state reaction.

In order that we may theoretically calculate the rate of formation of a spinel we must first make a few assumptions. These are: 1. The densities of the reactant oxides AO and B_2O_3 and of the reaction product are equal to their theoretical densities. That is, they are completely compact. Furthermore, the contact between them is ideal. 2. In the quasi-binary system AO-B_2O_3, only one compound, the spinel AB_2O_4, occurs. The ranges of homogeneity of all compounds are assumed to be negligibly small. 3. The reaction takes place isothermally. 4. Local thermodynamic equilibrium is maintained within the reaction layer and also at the phase boundaries. Under these assumptions, it is possible to calculate the rate of reaction, provided that the standard free energy of formation of the spinel and the component diffusion coefficients of the ions taking part in the reaction are known.

In Fig. 6-3 is shown a number of possible reaction mechanisms. Included here are mechanisms in which the oxygen is transported through the gas phase and local electrical neutrality is maintained by means of electronic semiconduction in the reaction product.

In practice, one can say that the transport coefficients of the individual ions are generally rather different from one another. In oxide spinels, the diffusion of oxygen is negligible compared to the cationic diffusion. Therefore, we can eliminate a number of the mechanisms shown in Fig. 6-3. Furthermore, if ideal contact is maintained at the phase boundaries so that the gas phase cannot enter, then the only remaining probable reaction mechanism is the counter-diffusion of cations. In this mechanism, the two cation fluxes in the reaction product are coupled through the condition of electroneutrality.

The spinel formation reaction $AO + B_2O_3 = AB_2O_4$ is a heterogeneous reaction. Therefore, material must be transported across the phase boundaries and through the reaction product. As the thickness of the reaction layer increases, the resistance to diffusion of the

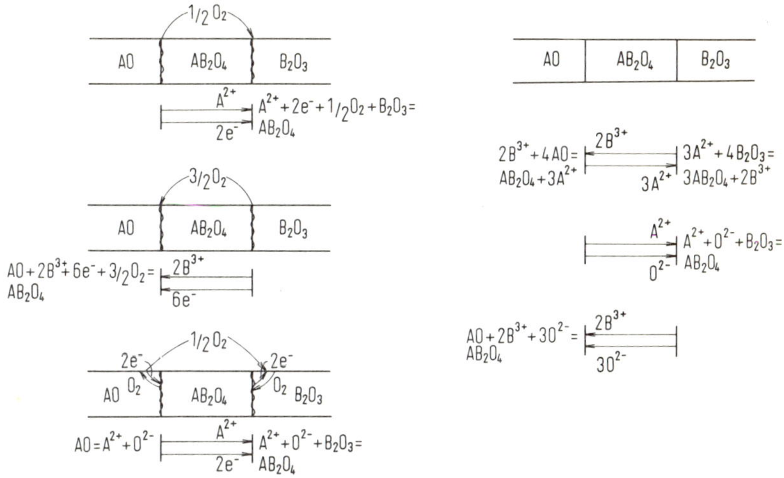

Fig. 6-3. Reaction mechanisms and corresponding phase boundary reactions for the spinel formation reaction $AO+B_2O_3=AB_2O_4$.

reaction product increases relative to the resistance of the phase boundaries. Finally, if the phase boundary resistance can be neglected relative to the diffusional resistance, then local thermodynamic equilibrium will be attained at the phase boundaries. In this case, there will no longer be a jump in the component activities at the phase boundaries, but only a break. For oxide spinel formation, this state is reached when the reaction layer has attained a thickness Δx of the order of 1 μ at the usual reaction temperature above 1200 °C. A parabolic rate law is now experimentally observed, and this can easily be explained as follows. First of all, we may assume that local equilibrium occurs at the phase boundaries. This means that, at the phase boundaries, all thermodynamic variables are fixed, and so the local defect concentrations are also fixed for all time. Therefore, an average concentration gradient of defects is formed in the reaction layer which is inversely proportional to the layer thickness Δx. This results in a flux of ions. The rate-determining ionic flux is therefore given by $j_i \propto 1/\Delta x$. Since j_i is proportional to the instantaneous growth rate $\mathrm{d}\Delta x/\mathrm{d}t$ of the layer, it follows that $\mathrm{d}\Delta x/\mathrm{d}t \propto 1/\Delta x$. This expression may be integrated to give the parabolic growth law in the form:

$$\Delta x^2 = 2\bar{k}t$$

$\bar{k}$ is known as the practical reaction rate constant. The parabolic growth of $NiAl_2O_4$ from NiO and Al_2O_3 is shown graphically in Fig. 6-4.

In order to treat the formation of spinels quantitatively, we begin with the flux equations for the various types of ions (cf. eq. (5-13)). The electrical potential ϕ is eliminated by means of the condition of electroneutrality, and the following equilibrium conditions:

$$A^{2+} + O^{2-} = AO; \qquad \eta_{A_2^+} + \eta_{O^{2-}} = \mu_{AO} \tag{6-19}$$

$$2B^{3+} + 3O^{2-} = B_2O_3; \qquad 2\eta_{B_3^+} + 3\eta_{O^{2-}} = \mu_{B_2O_3} \tag{6-20}$$

$$\mu_{AO} + \mu_{B_2O_2} = \mu^0_{AB_2O_4}; \qquad a_{AO} \cdot a_{B_2O_3} = \exp\frac{\Delta G^0_{AB_2O_4}}{RT} \tag{6-21}$$

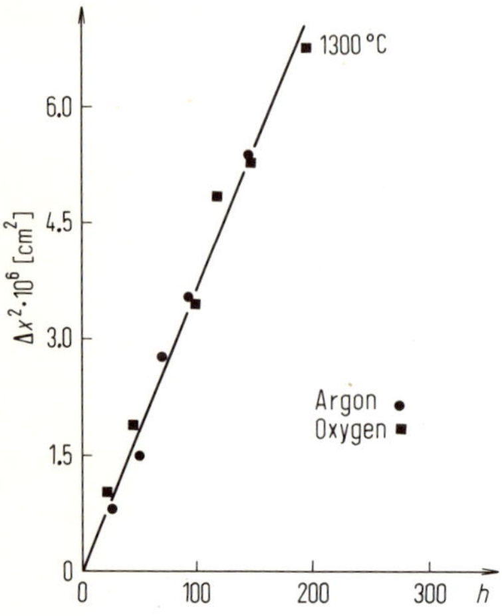

Fig. 6-4. The increase in the reaction layer thickness as a function of time during the formation of $NiAl_2O_4$ from NiO and Al_2O_3 ($T = 1300$ °C) according to [32].

We next perform a simple integration over the reaction layer, assuming that the ionic fluxes are constant across the reaction product. This assumption will be valid provided that the range of homogeneity of the reaction product is sufficiently narrow. We thus arrive, after a short calculation, at the parabolic rate law for the growth of the reaction layer AB_2O_4 in the form:

$$\frac{d\Delta x}{dt} = \frac{vk}{\Delta x}; \quad \Delta x^2 = 2vkt = 2\bar{k}t \tag{6-22}$$

Δx is the thickness of the layer at time t, k is the so-called rational reaction rate constant, and v is the increase in volume of the reaction layer resulting from the transport of one equivalent of ions through the reaction product. This volume increase can be calculated if we know the reaction mechanism and the molar volume of the reaction product. In order to calculate the reaction rate, then, it is necessary to calculate the rational reaction rate constant. By means of the calculation which was described above, the following expression is obtained for k:

$$k = \frac{\gamma}{RT} \int_{\mu'_{AO}}^{\mu''_{AO}} z_i c_i D_i \, d\mu_{AO} \tag{6-23}$$

where D_i is the component diffusion coefficient of the ions of sort i that are rate-determining. This equation will be valid provided that $D_1 \ll D_i \ll D_2$, where i, 1, and 2 are the three types of ions. In words, this means that the slowest ion plays no part in the reaction, while the movement of the fastest ion is coupled to the movement of the ion of intermediate velocity in order that electroneutrality be preserved. Therefore, it is the ion of intermediate velocity which determines the rate of reaction. γ is a numerical factor of the order of 1 which follows from the calculation. μ'_{AO} and μ''_{AO} are the chemical potentials of AO at either side of the reaction layer. Their difference is equal to the standard free energy of formation of the reaction product. From eq. (6-23) it can be seen that in order to calculate k exactly it is necessary to know the component diffusion coefficient as a function of the chemical potential of AO or B_2O_3. It has been shown in section 4.2.2 that the defect concentration (d) in the ternary crystal depends upon the chemical potential (or the activity) of a binary component according to the formula:

$$(\mathrm{d}) = (\mathrm{d})_{a_{AO}=1} \cdot a_{AO}^{n} = (\mathrm{d})_{a_{AO}=1} \cdot \exp \frac{n\,(\mu_{AO} - \mu_{AO}^{\circ})}{RT} \tag{6-24}$$

In particular, such a relationship applies to the concentrations of those defects (vacancies and interstitial ions) which are responsible for the diffusion in the reaction product. Therefore, the component diffusion coefficient D_i varies in the same way as the defect concentration in eq. (6-24). The factor n can be either positive or negative, and, as shown in section 4.2.2, it will have a value characteristic of the particular disorder type. Now, under the assumptions which have been made, the concentration c_i of ions of type i is constant, and the component diffusion coefficient D_i is known as a function of the chemical potential. Thus, we can substitute eq. (6-24) into eq. (6-23) and evaluate the integral with the result that:

$$k = \frac{\gamma z_i c_i D_i^{\circ}}{n} \left(1 - \exp \frac{n\,\Delta G^{0}_{AB_2O_4}}{RT}\right) \tag{6-25}$$

D_i° is the value of the component diffusion coefficient of the rate-determining ions of type i in the reaction product when the component activity $a_{AO} = 1$.

In polycrystalline or porous materials, the reaction rate can be influenced by diffusion along grain boundaries, inner surfaces, and dislocations, or across pores via the gas phase. Examples of this occur especially in the cases of ferrite or chromite formation.

Even if the reactants and the reaction product exhibit a wide range of homogeneity, which can no longer be neglected, it can be shown, by using the relationships which have already been discussed, that a parabolic growth law for the reaction layer still results, as long as the growth is one-dimensional and local equilibrium is maintained. Such a situation is realized for many spinels when the reaction temperature is high. The formation of $MgAl_2O_4$ according to the reaction: $MgO + Al_2O_3 = MgAl_2O_4$ is a good example. At 1800 °C, Al_2O_3 is soluble to 25 mole % in solid solution in $MgAl_2O_4$ [14]. In order to prove that the growth is parabolic, we substitute the dimensionless parameter $\xi\,(0\ldots1) = x/\Delta x$ into the flux equation of the rate-determining ion. Δx is the thickness of the reaction layer, and x is a distance coordinate which is zero at the phase boundary. The flux equation is then integrated over the reaction layer. Since the integration is performed between two fixed limits, it can be seen from the equation

$$\int_0^1 j_i \,\mathrm{d}\xi = -\frac{\gamma}{\Delta x} \int_{a_{AO}(\xi=0)}^{a_{AO}(\xi=1)} D_i c_i \,\mathrm{d}\ln a_{AO} = \frac{\text{const}}{\Delta x} \propto \frac{\mathrm{d}\Delta x}{\mathrm{d}t} \tag{6-26}$$

that a parabolic growth law is obeyed, since the left-hand-side of eq. (6-26) is an expression for the average flux in the reaction layer, and this average flux is proportional to the rate of increase of thickness $\mathrm{d}\Delta x/\mathrm{d}t$. The factor of proportionality includes the average molar volume of the reaction product. From the above treatment of the problem, it can be seen once again that it is only possible to treat the reaction quantitatively if the molar volume, the concentration, and the rate-determining component diffusion coefficient are known. However, for systems with wide ranges of homogeneity, the laws of ideal dilute solutions can no longer be applied to the defects as they were in the case of systems with narrow ranges of homogeneity. Eq. (6-24) is thus no longer applicable. As yet, no complete theoretical treatment of this problem has been possible.

In Fig. 6-5, experimentally measured rate constants for the formation of $NiGa_2O_4$ from NiO and Ga_2O_3 are compared with theoretical values calculated from eq. (6-25). The calculations were made with the assumption that the diffusion of cations is rate-determining. As can be seen from the plot, at sufficiently high temperatures where bulk diffusion prevails, the theoretical values of the rate constant are in good agreement with the experimental values.

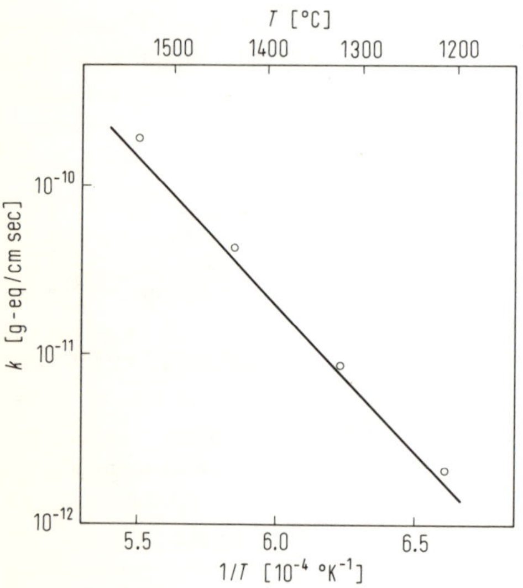

Fig. 6-5. Comparison of experimental and calculated rate constants for the formation of $NiGa_2O_4$ from NiO and Ga_2O_3 in air [43].

6.2.2. Silicate formation reactions

One need only reflect upon the importance of silicates in the refractories industry today to understand the interest in silicate formation reactions. The formation of calcium disilicate during the production of cement may also be mentioned here as a further important example of this type of reaction. In contrast to the case of spinel formation, however, there have been relatively very few fundamental studies of the mechanism of formation of silicates. The reasons for this are presumably that, firstly, reaction rates are generally very low; secondly, there are no suitable radioactive isotopes which can be used as radioactive tracers in diffusion experiments, and thirdly, silicates exhibit a very large variety of different structures. The question of the mechanism of formation of simple silicates from SiO_2 and binary oxides is analogous to the question of the formation of spinels. Structurally, however, the situation is different because the silicate is composed of $(SiO_4)^{4-}$ anion groups which can be joined together in many different ways. The cations are embedded among these anion groups. It may be assumed that, because of its size, an anion group cannot move as a unit through the crystal lattice. Also, it would seem most likely that the most mobile ions would be the cations which are embedded among the silicate groups, and therefore these are not the rate-determining ions. Furthermore, the bonding between the individual ions in the anion groups is very strong because of the partly covalent nature of these bonds. Thus, the question regarding the mechanism of silicate formation finally reduces to the question of whether oxygen ions or silicon ions (as rate-determining ions) diffuse faster in the silicate lattice. These conjectures are supported by measurements of cation diffusion in glasses and in simple silicates with the olivine structure where it has been found that the mobilities of those cations which are not bonded in the anion groups far exceed the mobility of Si^{4+} or of O^{2-} [15,16].

In the discussion of spinel reactions, it was stated that the fastest cations do not determine the reaction rate. The reason for this is the coupling between ionic fluxes in the reaction product which must occur in order to maintain electroneutrality. The reaction rate is actually determined essentially by that ionic species whose mobility is intermediate between that of the other two. Therefore it can be stated that in general the reaction rate will be determined by the silicon ions if they diffuse faster than the oxygen ions, and vice versa – as long as transport via the gas phase is excluded. There is as yet, however, an almost total lack of experimental information as to which of these two ions is rate-determining during the formation of silicates. From the results of interdiffusion experiments between silicates and germanates of the same structure it has been inferred indirectly that the orders of magnitude of the silicon and oxygen ion mobilities in simple silicates are not very different. This means that a general rule for deciding which ion is rate-determining in silicates cannot be simply postulated. Depending upon the particular reaction, either the silicon ions or the oxygen ions may determine the reaction rate. If the mobilities or component diffusion coefficients of the ions in the reaction product are known, then the reaction rate constant can be calculated in exactly the same manner as has just been described for the case of spinel formation.

Silicates which contain transition metal cations exhibit electronic semiconduction. For example, in Co_2SiO_4, p-type conduction is observed, and this is dependent upon the oxygen partial pressure. From the results of Haul in quartz glass [17], the occurrence of atomic diffusion of oxygen may be conjectured, and a possible formation mechanism involving the transport of atomic oxygen and electrons (or electron holes) as shown in Fig. 6-6 cannot be ruled out. Also, it may be anticipated, that the motion of silicon and oxygen ions in silicates is not uncorrelated. A correlation would explain the fact that their mobilities are found to have the same order of magnitude, although these ions are in no way comparable.

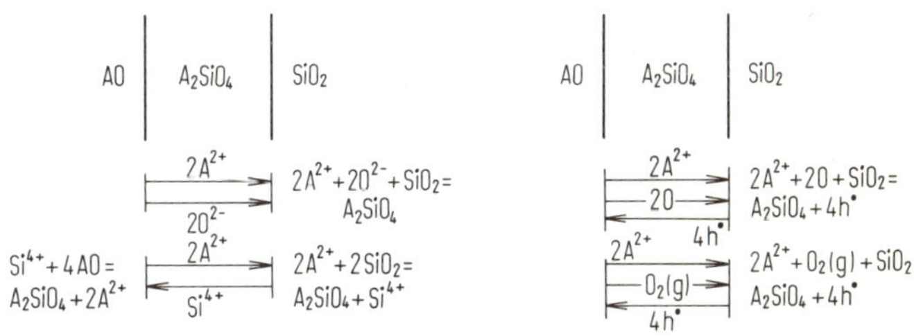

Fig. 6-6. Reaction mechanisms for the formation of simple orthosilicates.

6.2.3. Multiphase reaction products

If several compounds occur in the phase diagram between the quasi-binary reactants (for example, between the oxides AO and B_2O_3), then a multiphase product layer will be formed during the solid state reaction. This situation is shown in Fig. 6-7 where the activity of one component is plotted.

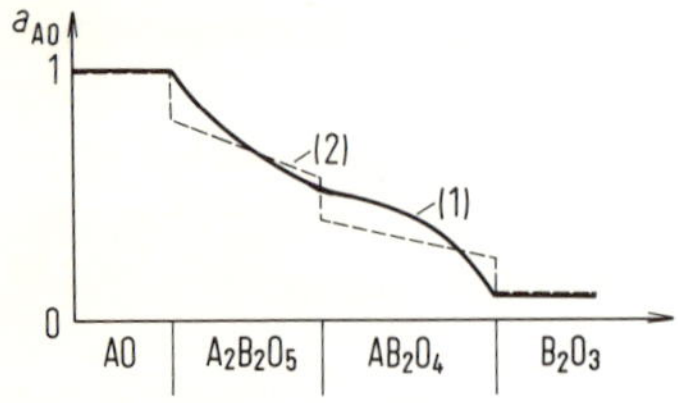

Fig. 6-7. Schematic plot of the activity of a reactant in a multiphase reaction product layer under the assumption that (1) local thermodynamic equilibrium is maintained, (2) local thermodynamic equilibrium is not maintained.

In the ternary system, the values of all thermodynamic variables will be fixed on all phase boundaries if local equilibrium is maintained and if the partial pressure of oxygen, as well as P and T, is fixed. Therefore, the defect gradient, and thus the particle flux, in every phase of the reaction product will be inversely proportional to the corresponding thickness of the phase. This, then, results in a parabolic growth law for each phase of the reaction product, as follows from the discussion of section 6.2.1. Let (p) be an index which denotes the phase, and let $\bar{k}$ be the practical rate constant. Then:

$$\Delta x^{(p)} = (2\bar{k}^{(p)} \cdot t)^{1/2}$$
$$\sum_p \Delta x^{(p)} = \sum (2\bar{k}^{(p)} \cdot t)^{1/2} = \sum (2\bar{k}^{(p)})^{1/2}\, t^{1/2} \tag{6-27}$$

Therefore, a parabolic growth law also applies to the total reaction layer thickness. Furthermore, since

$$\frac{\Delta x^{(p)}}{\Delta x^{(p')}} = \left(\frac{\bar{k}^{(p)}}{\bar{k}^{(p')}}\right)^{1/2} \tag{6-28}$$

it follows that the ratios of the thicknesses of the individual product phases are independent of time.

In order to make theoretical calculations, we assume first that the thermodynamics of the total quasi-binary system and the mobilities of the ions in the individual product phases are known. This means according to eq. (6-23) that the rational rate constants of the individual phases are known. If the rational rate constant of phase (p) is designated by $k^{(p)}$, then the differential equation for the increase in thickness of this phase is as follows:

$$\frac{\mathrm{d}\Delta x^{(p)}}{\mathrm{d}t} = \frac{k^{(p-1)} v_p^{(p-1)}}{\Delta x^{(p-1)}} + \frac{k^{(p)} v_p}{\Delta x^{(p)}} + \frac{k^{(p+1)} v_p^{(p+1)}}{\Delta x^{(p+1)}} \tag{6-29}$$

Such a differential equation may be written for each phase (p). In words, this equation says that if the reaction mechanism is known in each phase, then the increase in thickness of the p'th phase is given by taking the product of the ionic flux in this phase and the corresponding reaction volume v_p, and then subtracting from (or adding to) this term two additional terms to account for the transport occurring in the neighbouring phases $p - 1$ and $p + 1$. (Depending upon the reaction mechanism, $v_p^{(p\pm 1)}$ can be either positive or negative!) Because of the coupling of all these differential equations, it is obvious that the growth rate of the p'th phase, as described above by the parabolic rate constant $\bar{k}^{(p)}$, is dependent upon the thermodynamic and kinetic parameters of all the other phases which are being formed. A distinction can thus be made between rate constants of the first and second kind [18]. Rate constants of the first kind describe the growth of phase p when all phases $p = 1, 2, \ldots, n$ of the quasi-binary system are being formed simultaneously. Rate constants of the second kind describe the parabolic growth rate of the layer thickness when the phase p is formed only from the saturated phases $p - 1$ and $p + 1$. Finally, reference may be made here to the work of Kidson [36] on the formation of multiphase product layers.

Obviously, the reaction rate constant of the second kind is the more important one from a fundamental point of view. In a later section, it will be shown how to transform the reaction rate constant of first kind into second kind (and vice versa), if, generally speaking, the concentration profile of the reaction couple has been measured for a given reaction time.

Once again, it should be mentioned that the reaction products and the reactants together comprise a quasi-binary system which can be read from the phase diagram – for the case where local thermodynamic equilibrium is maintained at the phase boundaries. If local equilibrium at all the phase boundaries is not established, then it is possible that certain product phases do not form, even though they are present in the equilibrium phase diagram. The probability that local equilibrium at the phase boundaries will, in fact, occur becomes better and better as the product layer grows. Reactions as discussed here are most easily studied experimentally today by means of the electron probe (electron beam microanalysis).

6.2.4. Solutions of heterovalent components

This very important class of solid state reactions is frequently occurring together with the classical solid state reactions which have just been discussed. To be systematic, however, we should treat it in a separate section. In order to illustrate the general problem, let us consider the following experiment. $MgCr_2O_4$, which was prepared by the solid state reaction between MgO and Cr_2O_3, is brought into contact with a single crystal of MgO to give the sequence of layers $MgCr_2O_4/MgO$. For given values of P, T, and p_{O_2}, the activity of Cr_2O_3 at the phase boundary will be fixed. The numerical value of this activity is given as $a_{Cr_2O_3} = \exp(\Delta G^0_{MgCr_2O_4}/RT)$, where $\Delta G^0_{MgCr_2O_4}$ is the standard free energy of formation for the spinel. Now, Cr_2O_3 is soluble to a certain extent in MgO. The maximum solubility will occur when the MgO coexists with $MgCr_2O_4$, and this equilibrium value will first be reached at the phase boundary. Thereafter, chromium ions will diffuse into the MgO, according to Fick's second law with boundary conditions: $c_{Cr_2O_3}(x = 0) = c^\circ, t > 0$; $c_{Cr_2O_3}(x = \infty) = 0, t > 0$, where x is the spatial coordinate. The origin of the coordinate system is at the phase boundary. Analogous reactions are observed for the couples $NiAl_2O_4/NiO$, $MgFe_2O_4/MgO$ or K_2SrCl_4/KCl. These systems are representative of a large number of possible systems. The common factor in all these heterogeneous reactions is that the solute which dissolves in the binary crystal does so in the form of ions which possess a valence different from that of the cations of the matrix crystal. As a result, defects are formed in the matrix crystal in order to compensate for the differently charged ions. The local concentration of these defects depends upon the local concentration of the heterovalent solute. If, as in the present case, these defects are vacancies in the cation sublattice, then the concentration of these vacancies determines the diffusion coefficients of the diffusing components, since diffusion occurs via vacancies in the cation sublattice. In this case, then, the diffusion coefficient may vary directly with concentration: $\tilde{D} = D^0 \frac{c}{c^0}$. Fick's second law for this problem has been solved numerically by Wagner [19]. Recently, an analytical solution has been given [37]. Since $\tilde{D} \to 0$ as $c \to 0$, c will become zero at some finite point $x = x_+$. This permits the solution of the problem to be represented by a rapidly converging power series which, for a fixed phase boundary at $x = 0$ reads:

$$\frac{c}{c^\circ} = 1.306\,(z - \frac{1}{4}z^2 + \frac{1}{72}z^3 + \dots) \tag{6-30}$$

where $z = 1 - x/x_+$, and $x_+ = 0.8081\sqrt{4D^\circ t}$. A nearly linear concentration profile results as shown in Fig. 6-8, where the solution of Fick's second law for this case is compared to the solution when $\tilde{D}$ is constant. As can be seen, this curve can be approximated quite well by a linear function:

$$\frac{c}{c^\circ} = 1 - \frac{x}{0.8081\sqrt{4D^\circ t}} \tag{6-30a}$$

For solid-solid reactions, it is often the case that the phase boundary will not remain fixed at $x = 0$, but will move in accordance with a parabolic rate law. Eq. (6-30) can be modified to account for this [37]. As the velocity of the phase boundary becomes larger, higher terms in eq. (6-30) will become important, such that significant departures from a linear concentration profile are observed.

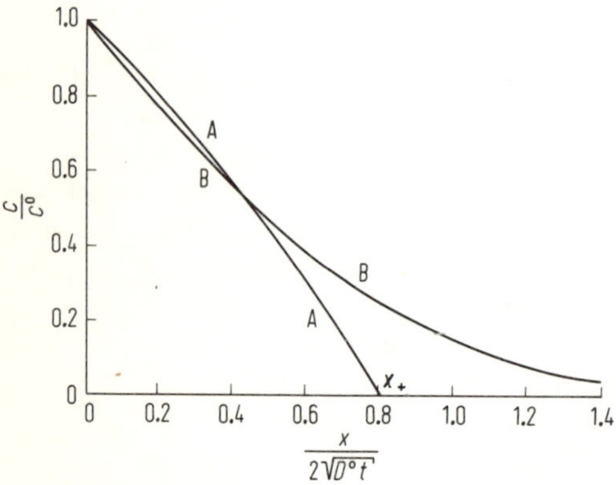

Fig. 6-8. The solution of Fick's second law for diffusion in a semi-infinite medium when the surface concentration c° is constant for the case of (A) a chemical diffusion coefficient $\tilde{D}$ of the form $\tilde{D} = D^\circ \frac{c}{c^\circ}$, and (B) $\tilde{D}$ = constant.

Reactions of this type occur very frequently. For instance, they occur during the formation of spinels or silicates whenever the reactants exhibit significant mutual solubility [20]. In such cases, the growth rate constants will be different from the rate constants when there is negligible terminal solubility, since in the former case a certain amount of the components will dissolve in the end phases during the course of the reaction. The parabolic growth law will still be obeyed, however. The details of this complicated kinetic process will be discussed in chapter 8 along with the analogous process of the solution of gases in metals during tarnishing. This type of reaction will also occur – albeit in the inverse sense – for the case where the heterovalent solute ions precipitate out of a supersaturated matrix of a binary crystal. Such a situation is frequently observed when the temperature is decreased following the establishment of equilibrium at higher temperatures [21].

6.2.5. Phase boundary reactions

During reactions involving the formation and growth of new phases, there will always be a flux of matter across the phase boundaries. When atomic particles attempt to cross a phase boundary they encounter a resistance. Therefore, in order that a certain particle flux be maintained, a discontinuity in the activity of the components in question must occur. As the thickness of the product layer increases, the diffusional resistance of this layer continually increases relative to the phase boundary resistance. The particle flux and the activity gradient in the phase boundary both become smaller until, finally, a state of local equilibrium is achieved at the phase boundary, and the discontinuity in activity virtually disappears. For very small product layer thicknesses, on the other hand, the diffusional resistance of the product layer is negligible, and the discontinuity in the activity at the phase boundary takes on the maximum value which is thermodynamically possible. In this case, the reaction rate is completely determined by the resistance of the phase boundary. The structure of the phase boundary is determined by the two bordering phases. Since these phases do not change during the reaction, except to alter their dimensions, it can be seen that the resistance of the phase boundary to the passage of particles is constant. This means that a linear growth law will be obeyed as long as the phase boundary reaction is rate-determining. Such linear growth is actually observed during the formation of spinels and silicates at the beginning of the solid state reactions. An example is given in Fig. 6-9.

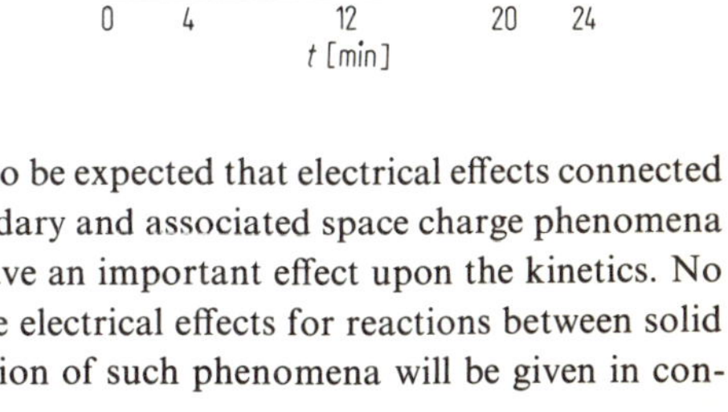

Fig. 6-9. The linear variation with time of the increase in mass Δm as well as the increase in thickness Δx during the formation of Zn_2SiO_4 from ZnO and cristobalite in air at 1 348 °C.

For very thin reaction product layers, it is to be expected that electrical effects connected with the electrical double layer at the phase boundary and associated space charge phenomena in the vicinity of the phase boundary will also have an important effect upon the kinetics. No information is presently available regarding these electrical effects for reactions between solid phases. However, in chapter 8 a detailed discussion of such phenomena will be given in connection with another class of solid state reactions.

Very little is known about the atomic mechanism of phase boundary reactions. If ideal contact between the reactant and the reaction product can be assumed, then we can distinguish the following limiting cases. 1. One or more sublattices of the reactant remain unchanged during the reaction. An example would be the boundary $MgO/MgCr_2O_4$ in the reaction $MgO + Cr_2O_3 = MgCr_2O_4$. In both the reactants and the reaction product, the large oxygen anions are found in the same face-centred-cubic close-packed sublattice. The phase boundary reaction simply amounts to a redistribution of cations on the octahedral and tetrahedral lattice sites. This also explains the strict epitaxy between the reactant and the product. 2. One or more sublattices are converted into one another by means of cooperative processes at the phase boundary. An example is the boundary $Al_2O_3/CoAl_2O_4$ during the reaction $CoO + Al_2O_3 = CoAl_2O_4$. Here, the hexagonal-close-packing of the large oxygen ions in the α-corundum can be converted to the cubic packing of the spinel through the motion of partial dislocations [22]. At the same time, a redistribution of cations takes place at the phase boundary. As long as the dislocations are not retarded in their movement by impurities, inclusions, etc., the cation redistribution along with their diffusion into the reactant (which results in a certain degree of supersaturation) will determine the phase boundary reaction rate. Once again, we expect strict epitaxy. In the present case, the hexagonal c-axis $\langle 000.1 \rangle$ of corundum should be parallel to the cube diagonal $\langle 111 \rangle$ of the spinel in order that the close-packed planes of oxygen ions remain parallel. This is, in fact, observed. 3. The structures of the reactants and the reaction product are not at all related to each other. The phase boundary is then a highly disordered region which may be compared with grain boundaries or with a thin lamella of liquid. Epitaxy may still perhaps be expected as a result of epitaxial nucleation of the new phase which is forming. Examples are the reactions between CoO and TiO_2, or between PbO and SiO_2 [39].

Finally, it must be noted that ideal contact between the various phases is not always achieved. If ideal contact does not occur, then the reaction will be affected by gas transport across narrow gaps and fissures. The rate-determining step of the phase boundary reaction could then be evaporation, condensation, or gas transport.

6.3. Reactions between single crystals and powder reactions

In the experimental study of elementary reaction mechanisms, it is advantageous to use single crystals as far as is possible (if they are available) since, in general, the boundary conditions are thereby simplified. The situation is entirely different, however, in the case of powder reactions, which are technologically so important. Depending upon the degree of approximation required in a quantitative treatment, we must be concerned with the following parameters: grain size, grain size distribution, grain shape, contact area between the grains as a function of time, vapour pressure, and rate of vaporization. One can easily see that a complete description of the kinetics of a powder reaction would have to take into account so many parameters, and would thus be so complicated, that it would be virtually impossible to give a fundamental interpretation to the reaction constants. Therefore, in the following discussion, these two types of reactions will be treated separately.

6.3.1. Reactions between single crystals

By means of a suitable geometrical configuration of the reactants, we can ensure that we have a one-dimensional reaction problem. It is then possible to reach conclusions about the reaction mechanism from the morphology of the reaction product for the case where the reactants and products have a common sublattice, or for the case where certain sublattices become interconverted through cooperative processes. Let us take as example the reaction: $NiO(SC) + Al_2O_3(SC) = NiAl_2O_4(SC)$. (SC = single crystal). Assume that ideal contact occurs at the phase boundaries, and that the reaction mechanism is the counterdiffusion of cations. We then expect that on the NiO side the reaction product will retain the oxygen sublattice and the orientation of NiO. That is, $(100)_{NiO} = (100)_{spinel}$, etc. On the Al_2O_3 side, the reaction product will have an orientation given by $\langle 111 \rangle_{spinel} = \langle 000.1 \rangle_{\alpha-Al_2O_3}$. This results from the motion of partial dislocations which transform the α-corundum lattice into the spinel lattice. Furthermore, the fraction of reaction product formed at the phase boundary spinel/NiO and the fraction formed at the boundary spinel/Al_2O_3 are in the ratio 1 : 3 as can be seen from the reaction mechanism as discussed in section 6.2.1. Thus, the morphology, as illustrated in Fig. 2-1, page 14, and Fig. 6-3, page 91, reflects both the phase boundary reaction and the counterdiffusion of cations in the reaction product. Analogous examples may be found for other reactions and situations.

6.3.2. Powder reactions

Traditional ceramic processes for the preparation of solid products consist of the mixing and firing of powdered reactants. For example, titanate dielectrics and ferrite control or circuit elements are produced in this way. The art of the ceramist consists in properly choosing the grain size and the temperature programming so as to obtain the optimal structure. A frequently used kinetic equation for powder reactions has been given by Jander [23]. In this equation, oversimplified assumptions have been made. The derivation of the equation is based on the premise that equal-sized spheres of reactant A are embedded in a quasi-continuous medium of reactant B. A parabolic growth law is assumed for the increase in thickness of the spherical shell of reaction product P which is forming around the A particles. If x is the relative amount of A transformed into reaction product, and r_A is the radius of the original spherical particles of A, it then follows that:

$$r_A^2 \, [1 - (1 - x)^{1/3}]^2 = 2 \bar{k} t \tag{6-31}$$

where $\bar{k}$ is the practical parabolic rate constant. Two fundamental points have been neglected in this derivation, and these must be corrected: 1. Parabolic growth is expected for a diffusion-controlled one-dimensional reaction problem, but not for a problem with spherical symmetry. In the latter case, a parabolic growth law will describe, at best, only the initial stages of the reaction when the product thickness $\Delta x \ll r_A$. 2. In eq. (6-31), the difference in molar volume between reactants and reaction product has not been taken into account. Finally, the assumption that the reactant B completely envelops the A particles only comes close to being valid for extremely large values of the ratio r_A/r_B. The calculation has been repeated by Carter [24] who took these objections into account and thereby obtained the following equation for the time dependence of the relative extent, x, of the reaction:

$$[1 + (z - 1)\, x]^{2/3} + (z - 1)\,(1 - x)^{2/3} = z + 2\,(1 - z)\frac{\bar{k} t}{r_A^2} \tag{6-32}$$

where z is the volume of the reaction product formed from a unit volume of reactant A. The experimental results shown in Fig. 6-10 confirm that under the given assumptions eq. (6-32) is applicable to the reaction $ZnO + Al_2O_3 = ZnAl_2O_4$ for extents of reaction up to 100%.

In this experiment for the formation of zinc aluminate, the prerequisites were fulfilled by suspending a number of equal-sized spheres of α-corundum in a ZnO crucible and bringing the system to the reaction temperature. The vapour pressure of ZnO is large enough to ensure that the activity $a_{ZnO} = 1$ at the surface of the spheres during the entire course of the reaction. We wish now to compare this gas-solid reaction with a classical solid state powder reaction. However, in the latter case we should expect that growth of the reaction product will only occur at contact points, and, for more or less equal-sized particles of the reactants, the number of such contact points is very small. Nevertheless, it is observed that in many cases the Jander equation (6-31) or the Carter equation (6-32) gives a satisfactory description of at least the initial stages of the reaction. Thus, it must be assumed either that rapid surface diffusion provides a uniform supply of one of the reactants over the entire surface of the other, or that the vapour pressure and the rate of vaporization of one of the reactants is high enough so that many powder reactions are not actually classical solid state reactions, but rather, they are gas-solid reactions. The latter conjecture has been verified in an experiment in which a thermobalance was used to measure the increase in weight of a single crystal of Al_2O_3 which was suspended inside a ZnO crucible such that the reaction could occur only via the gas phase. The rate

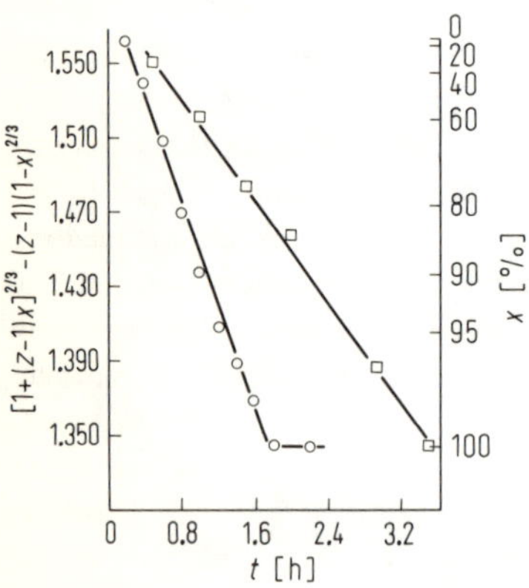

Fig. 6-10. Proof of the Carter equation (6-32) for the reaction of spherical particles in the reaction $ZnO + Al_2O_3 = ZnAl_2O_4$ at 1400 °C in air. r_A (○) = 19 microns; r_A (□) = 25 microns.

constant for the reaction $ZnO(g) + Al_2O_3(s) = ZnAl_2O_4(s)$ which was measured in this manner is in excellent agreement with the rate constant measured for the classical solid state reaction $ZnO(s) + Al_2O_3(s) = ZnAl_2O_4(s)$. If it cannot be assumed that a constant activity of one reactant at the surface will be maintained, either through high surface diffusion or through transport via the gas phase, then the reaction kinetics will be influenced not only by the ratio r_A/r_B, but also by the absolute values of r_A and r_B. The reason for this is that beyond a certain value of r_A the particles of reactant A, which are assumed to be spherical, come into contact with each other. At the points of contact, reaction with the reactant B is, of course, no longer possible.

In the literature [6] one can find relationships which allow such parameters as contact probability, grain size distribution functions, and deviations from spherical symmetry to be taken into account in calculations involving powder reactions. How well such expressions describe the actual reality of powder reactions, however, remains an open question, since they neglect surface diffusion and gas transport.

6.4. Displacement reactions

The so-called displacement reaction AX + BY = AY + BX is a possible limiting case of the chemical reaction between two reactants AX and BY with different cations and different anions. Depending upon the ionic mobilities and the miscibilities, very different types of reactions between AX and BY can occur. Continuing the theme of the sections 6.2 and 6.3 which dealt with heterogeneous reactions, we shall only discuss the limiting case of the displacement reaction AX + BY = AY + BX when the phases AX, BY, AY, and BX exhibit negligible mutual solubility. Furthermore, we shall assume that the cations A and B are far more mobile than the anions X and Y. This is most frequently the case in practice. We can then distinguish between two limiting mechanisms. These mechanisms, which were proposed by Jost [7] and Wagner [25], are illustrated in Fig. 6-11.

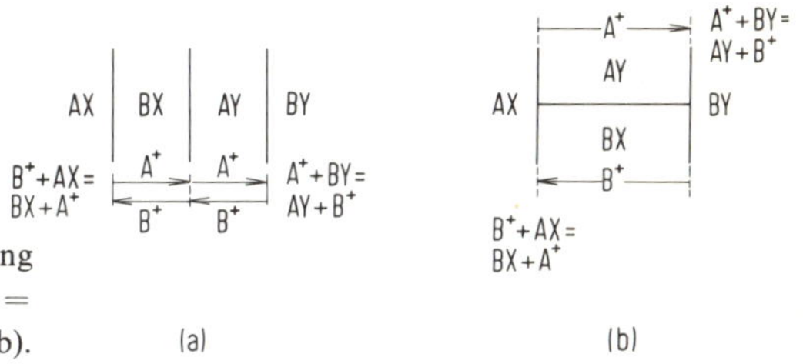

Fig. 6-11. Reaction mechanisms for two limiting cases of the displacement reaction AX + BY = = AY + BX according to Jost (a) and Wagner (b).

According to Jost, the reactants AX and BY are separated from one another by the product phases AY and BX. A layer of BX grows to cover the reactant AX, and a layer of AY grows to cover the reactant BY. This situation results from the fact that only the cations are mobile. In order that the reaction can proceed, it is necessary that there be a slight solubility of A^+ ions in BX and of B^+ ions in AY, since these ions must diffuse through the product phases. In order that this problem may be treated quantitatively, it is, in principle, necessary that the partial pressures of X_2 and Y_2 in the surrounding gas phase, as well as P and T, be fixed. The activity gradients of the cations in the product phases are then uniquely determined, and if the transport coefficients are known, the increase in thickness can be easily calculated, since the cation fluxes at the phase boundary BX/AY must be coupled.

Wagner begins with the assumption that in certain cases the solubilities and mobilities of A^+ ions in BX and of B^+ ions in AY are so low, and consequently the product layers are formed so slowly, that another reaction mechanism may predominate once nucleation has occurred. This reaction mechanism is shown in Fig. 6-11 (b). Essentially, a closed circular flow of cations occurs in the product phases such that the cations diffuse only in their own respective compounds. A quantitative approximation to the reaction rate can be made as follows. Since three phases are in simultaneous contact, it is sufficient to specify only one additional variable in addition to P and T in order to uniquely determine the problem. If the partial pressure p_{X_2} is chosen as this variable, then it is a simple matter to calculate the activity gradient of A in AY if the free energies of formation of the individual compounds are known. This is essentially given by the standard free energy ΔG^0 of the reaction AX + BY = AY + BX. Then, if the diffusional resistances in AY and BX are known, it is possible to calculate the rate of the displacement reaction for this limiting case as well.

In metal displacement reactions, i.e. $Fe + Cu_2O = FeO + 2Cu$, one often finds a reaction product morphology as predicted by the Wagner mechanism. This means that, instead of Jost's layer structure, one finds at a given coordinate perpendicular to the original interface a two phase mixture in which both reaction products occur simultaneously. In recent studies on metal displacement reactions between metals and oxides, criteria have been given for the occurrence of Jost's mechanism with a planar phase boundary between the two reaction product layers. For example when Cu (more noble metal) is the metallic reaction product, if the transport coefficient of oxygen in this metallic product layer is greater than the transport coefficient of the reactant metal (less noble metal) in the product oxide (Co in CoO), then planar phase boundaries are found which are stable under growth conditions. Thus, in complete analogy to metal oxidation in tarnishing reactions (see chapter 8), the oxide of the less noble metal can grow, since oxygen with the high activity of the two phase mixture (Cu, Cu_2O) of the more noble metal and metal oxide is supplied in sufficient amount by diffusion through the more noble metal according to the above assumption. If, on the other hand, there is a deficiency of oxygen at this phase boundary, then the less noble metal will seek out the oxygen supply as far as is possible, since the transport coefficient of the less noble metal in the oxide layer is greater than the oxygen transport coefficient in the product metal. In this case, according to quite general principles, one would expect nonplanar unstable growing phase boundaries. As a result, two phase reaction products are found in accordance with the morphology of the Wagner mechanism. A quantitative treatment of this latter kind of displacement reaction and an expression for the reaction rate can be found in the recent literature [40]. The principle of a maximum reaction rate in this process has been used in the theoretical treatment.

Two reactions of this kind, which were studied many years ago by Tammann as powder reactions, and which have more recently been extensively examined as reactions between single crystals and dense pressed sinters, are [26, 27]:

$$PbS + CdO = PbO + CdS \quad \text{and} \quad ZnS + CdO = ZnO + CdS$$

Electron probe studies have shown that the Jost mechanism is most likely in these cases. Other reactions of the general type discussed here which have been experimentally studied are the displacement reactions:

$$Cu + AgCl = Ag + CuCl \quad \text{and} \quad Co + Cu_2O = 2\,Cu + CoO$$

A complicated situation arises in the case of displacement reactions in which the reactants are heterovalent, since dopant effects can have a marked influence on the transport coefficients in the product phases. An example is the reaction:

$$PbCl_2 + 2\,AgI = PbI_2 + 2\,AgCl$$

In the technically important reaction:

$$Cu_2S + 2\,Cu_2O = 6\,Cu + SO_2\,(g)$$

one product phase is a gas. The morphology of this reaction is schematically shown in Fig. 6-12. It can be seen that, in this solid state reaction, the various partial reaction steps of the overall reaction occur at different places. An analogy can be drawn with the principle of the spatial separation of the different reaction steps in the electrochemistry of aqueous solutions [28].

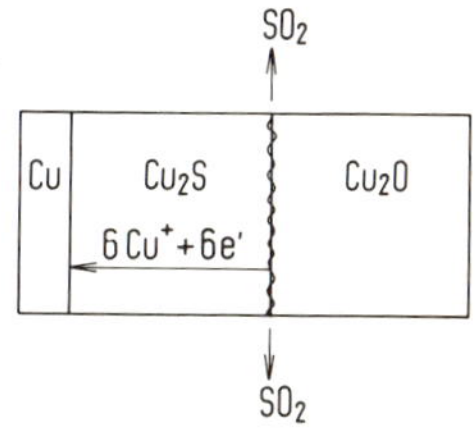

Fig. 6-12. Schematic diagram of the morphology of the exchange reaction $Cu_2S + 2\,Cu_2O = 6\,Cu + SO_2$ (g).

In the present case, the metallic reaction product does not separate the reactants, and so, again, a maximum reaction rate is possible.

6.5. Transport in glass

Because of its technological importance, and in connection with the discussion of silicate reactions in section 6.2.2, some fundamental ideas are presented here concerning material transport in glass.

To a physical chemist, glasses are frozen-in supercooled liquids. The term "frozen-in" means that the relaxation time for a cooperative translational movement has become quite large compared with the time for an elementary diffusional step of ions in glass.

If we limit ourselves to the important group of silica-containing glasses, the elementary structural units of which consist of SiO_4^{4-}-tetrahedra which are surrounded by cations like Na^+ (a network modifier, in contrast to network former cations like Si^{4+}) and under certain conditions by free oxygen ions, then diffraction methods [42] clearly show that these glasses do not exhibit long range order. Rather, they exhibit a pronounced short range order, just as is found in liquids. Depending upon the concentration ratio of network-forming oxides to network-modifying oxides, the SiO_4^{4-}-tetrahedra form more or less extended network regions (polyanions) in which the corners of the tetrahedra are formed by bridging oxygen ions between the tetrahedra. If the ratio of network-forming oxides to network modifiers becomes too small (i.e. if the silica tetrahedra become essentially isolated from each other), then there is no longer a tendency to form a glass [41].

Fig. 6-13 shows the glass structure in a schematic sketch. Since, in contrast to crystals, glasses have no long-range order, but are structurally disordered, there is always the possibility of elementary diffusional steps of the ionic constituents. It is rather easy to understand (see Fig. 6-13) the mobility of the network modifying ions. Usually, their mobility is far greater than the mobility of silicon ions or oxygen ions, since oxygen ions are mainly found in glasses either as bridging ions or as nonbridging oxygen ions which are nevertheless relatively strongly bonded to the silicon ions.

By comparing mobilities obtained from tracer measurements with those from electrical conductivity measurements for network-modifying ions with known transference numbers, one can calculate correlation factors f. These are of the order of 1/3 to 1/2. Conclusions, drawn from the magnitude of these correlation factors, concerning the mechanism of diffusion

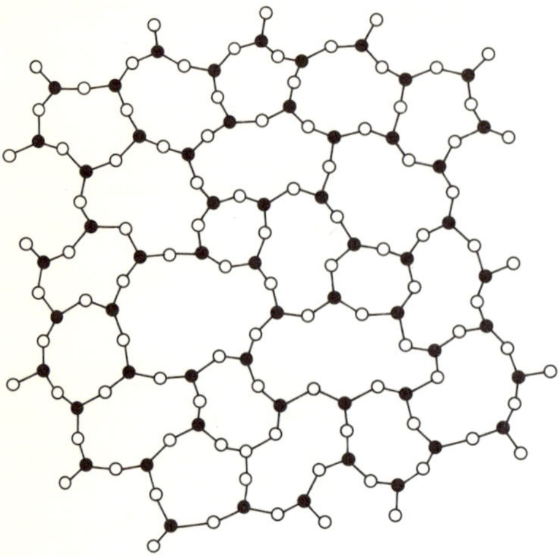

Fig. 6-13. Schematic two-dimensional representation of a glass network. The fourth bond of the Si ions is directed out of the plane. ● = Si^{4+}; ○ = O^{2-}.

in glass have to be regarded with some reservation. In contrast to diffusion in ionic crystals, no discrete point defects take part in the diffusion process in glass. The occurrence of correlation during the thermal statistical movement of cations in glasses can simply result from the fact that, for example, (as can be seen in Fig. 6-13) an elementary step of a sodium ion will disturb the average local distribution of electrical charge somewhat. This then leads to an increase in the local electric field, and, by relaxation processes, this results in a correlation of diffusional steps. One can conclude from this discussion that the physical reason for the occurrence of a "memory effect" in glass diffusion is quite different in kind from the "memory" of a radioactive tracer which — for example — is moving through an ionic crystal by way of vacancy diffusion. By consulting eqs. (5-24) and (5-25) one may say that, in a zeroth approximation, sodium ions which have just hopped in sodium glass cancel 1/4 of their diffusional steps because of an electrical field which immediately pushes them back from whence they came.

Some numbers may illustrate the order of magnitude of diffusion coefficients which we expect in silica glass [15]. The diffusion coefficient of Na ions at 1000 °C amounts to about 10^{-5} cm^2/sec, while at the same temperature the diffusion coefficient of Ca^{2+} ions is found to be about 10^{-8} cm^2/sec (smaller because of its higher electrical charge). The oxygen ion diffusion coefficient under the same conditions has been found to be about 10^{-13} cm^2/sec. This small value is explained by the bridging function of oxygen between the silica tetrahedra. At 200 °C, the diffusion coefficient of sodium ions has about the same value as the oxygen ion diffusion coefficient at 1000 °C.

It is interesting to note that the diffusion coefficient of Al^{3+} ions at 1000 °C has approximately the same value as the oxygen diffusion coefficient. This means that Al^{3+} ions will function at least partly as network-forming ions in place of Si^{4+} ions, and are at least partially built into the oxygen tetrahedra. Again, this experimental fact should be seen in conjunction with experimental results which show that, in crystalline silicates, the mobilities of oxygen ions and of silicon ions are of comparable magnitude (see section 6.2.2).

At the end of this discussion on ion mobilities in glass, it should be mentioned that the concentration profiles from tracer diffusion experiments do not in general exhibit a simple error function behaviour, if the experimental conditions of Fig. 5-5 are obeyed. Since, in tracer experiments, the thermodynamic factor (see eq. (5-28)) is equal to one, and since, in a homogeneous material, the diffusion coefficient is constant, the conclusion which has been drawn from this experimental fact is that glasses are metastable with respect to decomposition into different phases. This means that, in submicroscopic regions, glass may well be either spinodally decomposed (see section 7.4) or may exist in a multiphase structure. If the different glass phases are homogeneous in themselves and are continuously interwoven, and if, furthermore, the tracer ions can cross the phase boundaries, then one can anticipate quite complex diffusion profiles whose shapes will depend upon the ratio of the different phases in the phase mixture and upon the ratio of the mobilities in the different phases. An exact mathematical solution of the problem cannot be given.

In principle, in glasses, as well as in crystalline materials, one must distinguish between the determination of ionic mobilities by tracer methods and the chemical diffusion in inhomogeneous phases where concentration gradients of the components occur. Systematically speaking, diffusion in silicate glass is classified, in general, with multicomponent ionic diffusion processes. The basic ideas on this have been discussed in section 6.1.3. A transition from crystalline to amorphous material is unimportant for the phenomenological treatment. The formal treatment of the problem of multicomponent (ionic) diffusion by Cooper [38] shows how one must proceed in order to calculate the component concentrations as functions of time and space coordinates if the initial and boundary conditions of the diffusion problem are given along with the individual ion mobilities and the thermodynamics of the glass system. The necessary thermodynamic parameters are known if $(n - 2)$ activity coefficients have been measured as functions of composition of the n-component glass system since two more conditions in the form of the Gibbs-Duhem-relation and the electroneutrality condition exist. As in the case of diffusion in ternary metallic systems which will be discussed in detail in section 7.1.3, one anticipates up-hill diffusion of one or several components at certain time intervals and space regions in glass systems as the result of electrical coupling of fluxes and the influence of the concentrations upon the activity coefficients. Analoguous processes are found in what is called the "active transport" in membranes of biological systems.

All these ideas are of great practical influence and importance in glass technology. For instance, the homogenization of a multicomponent glass during its fabrication is diffusion controlled. Ion exchange, spinodal decomposition, or precipitation of crystalline phases may also be mentioned as some further examples of technically important and diffusion controlled processes in glass.

6.6. Literature

General Literature:

[1] A.L. Stuyts et al., Mater. Sci. Eng. *3*, 317 (1968/69).
[2] A.A. Frost and R.G. Pearson, Kinetik und Mechanismen homogener chemischer Reaktionen, 2nd ed., Verlag Chemie GmbH, Weinheim 1964.
[3] H. Reiss, C.S. Fuller and F.J. Morin, Bell System Techn. J. *35*, 535 (1956).
[4] K. Hauffe, Reaktionen in und an festen Stoffen, 2nd ed., Springer-Verlag, Berlin 1966.
[5] H. Schmalzried, Ber. Dtsch. Keram. Ges. *42*, 11 (1965).
[6] G.M. Schwab in J.W. Mitchell, R.C. DeVries, R.W. Roberts and P. Cannon, Reactivity of Solids, Proc. 6. Intern. Symp. 1968, Wiley-Interscience, New York 1969, p. 163.
[7] W. Jost, Diffusion und chemische Reaktion in festen Stoffen, Steinkopff-Verlag, Dresden 1937, p. 180.

Special Literature:

[8] T.R. Waite, J. Chem. Phys. *32*, 21 (1960).
[9] W. Rogalla and H. Schmalzried, Ber. Bunsenges. phys. Chemie *72*, 12 (1968).
[10] P. Penning, Philips Res. Repts. *14*, 337 (1959).
[11] S.L. Blank and J.A. Pask, J. Amer. Ceram. Soc. *52*, 669 (1969).
[12] G. Zintl, Z. phys. Chem. NF *48*, 340 (1966).
[13] W. Müller and H. Schmalzried, Z. phys. Chem. NF *57*, 203 (1968).
[14] D.M. Roy, R. Roy and E.F. Osborn, J. Amer. Ceram. Soc. *36*, 149 (1953).
[15] G.H. Frischat, Habilitation thesis, Technical University, Clausthal, Germany 1970.
[16] G. Borchardt, Diploma thesis, Technical University, Clausthal, Germany 1968.
[17] R. Haul and G. Dümbgen, Ber. Bunsenges. phys. Chem. *66*, 636 (1962).
[18] C. Wagner, Acta Met. *17*, 99 (1969).
[19] C. Wagner, J. Chem. Phys. *18*, 1227 (1950).
[20] C.D. Greskovich and V.S. Stubican, J. Phys. Chem. Solids *30*, 909 (1969).
[21] W. Rogalla, Ber. Bunsenges. phys. Chem. *72*, 615 (1968).
[22] M.L. Kronberg, Acta Met. *5*, 507 (1957).
[23] W. Jander, Z. anorg. allg. Chem. *163*, 1 (1927).
[24] R.E. Carter, J. Chem. Phys. *34*, 2010 (1961), *35*, 1137 (1961).
[25] C. Wagner, Z. anorg. allg. Chem. *236*, 320 (1938).
[26] V. Leute, Z. phys. Chem. NF *59*, 91 (1968).
[27] G. Tammann, Z. anorg. allg. Chem. *149*, 21 (1925).
[28] C. Wagner and W. Traud, Z. Elektrochem. *44*, 391 (1938).
[29] J. Crank, The Mathematics of Diffusion, Clarendon Press, Oxford 1957.
[30] E. de Lamotte and J.S. Kirkaldy, Z. phys. Chem. NF *67*, 31 (1969).
[31] K. Natesan and W.O. Philbrook, Trans. AIME *245*, 1417 (1969).
[32] F.S. Pettit et al., J. Amer. Ceram. Soc. *49*, 199 (1966).
[33] J.S. Armijo, J. Oxidation of Metals *1*, 171 (1969).
[34] E. Aukrust and A. Muan, Trans. AIME *227*, 1378 (1963).
[35] R.P. Nelson, Acta Polytechn. Scand. *60*, 2 (1967).
[36] G.V. Kidson, J. Nucl. Mater. *3*, 21 (1960).
[37] A.D. Pelton and T.H. Etsell, Acta Met. *20*, 1269 (1972).
[38] P. K. Gupta and A.R. Cooper Jr., Physica *54*, 39 (1971).
[39] K. Hardel and B. Strocka, Z. phys. Chem. NF *83*, 234 (1973).
[40] G.J. Yurek, Doctoral Thesis, Ohio State University, Columbus 1972.
[41] H. Scholze: Glas, Friedr. Vieweg und Sohn, Braunschweig 1965.
[42] B. E. Warren, J. Amer. Ceram. Soc. *24*, 256 (1941).
[43] W. Laqua, Thesis, T. U. Berlin 1971.

7. Reactions in the solid state – metals

7.1. Reactions in single-phase systems

In many respects, reactions between metals are simpler than reactions between ionic crystals. However, the discussion of metal reactions has been left until now since the main emphasis so far has been on the more chemical aspects of solid state reactions, and the reactions between ionic crystals serve as classical examples for this.

In a systematic classification of chemical reactions, we should once again make a distinction between reactions in single-phase and in multiphase metallic systems. If we start our discussion with single-phase systems, then, as in the previous chapter, we must consider both homogeneous reactions and reactions in inhomogeneous single-phase systems. For the case of ionic crystals, we recall that for certain types of point defect disorder, equilibration can occur only by means of homogeneous defect reactions in the solid. In simple metals, on the other hand, this is not the case. Homogeneous reactions between defects are not as important here as in the case of ionic crystals (although the coming together of two simple vacancies to form a double vacancy is a homogeneous reaction, and the formation of this type of defect complex can often be observed in metals). The condition of electrical neutrality, which is so basic for ionic crystals, also does not arise explicitly in the case of metals. Therefore, in metals we find single defects rather than oppositely charged defect pairs. These single defects (vacancies and interstitial atoms) are formed at sites of repeatable growth in the crystal – that is, at dislocations, at low- and high-angle grain boundaries, and on surfaces – according to the reaction equations:

$$Me_{defect} + V_i = Me_i \quad \text{or} \quad Me_{Me} = Me_{defect} + V_{Me} \tag{7-1}$$

The subscript "defect" refers to the site of repeatable growth. These equations can be compared to eq. (4-19) which is applicable for a homogeneous defect reaction in ionic crystals, and to eq. (6-6) which is applicable for an inhomogeneous defect reaction in ionic crystals. In these equations, in contrast to eq. (7-1), the pairwise formation of defects is evident.

Accordingly, the simplest reaction which we can treat in homogeneous metallic systems is the equilibration of defects. After this comes the broad field of reactions in single-phase systems with spatially variable composition. These latter reactions are generally classified under the title "Diffusion in Metals". There is a number of excellent monographs on this topic [1, 2, 3, 4]. The diffusional mechanisms and the fundamental phenomenological laws have already been treated in the introductory chapters, and especially in chapter 5. However, there are still a great many aspects of chemical reactions between metals which have not yet been discussed. These will be treated in detail in the present chapter.

7.1.1. Equilibration of vacancies

Our first example of a reaction in inhomogeneous metallic systems will be the equilibration of vacancies. In close-packed elemental crystals such as Cu or Al, essentially the only point defects are vacancies – as long as the metal is sufficiently pure. According to Gibbs, the concentrations of these vacancies in thermodynamic equilibrium at normal pressure is a function solely of temperature. Assume that a cylindrical metal whisker is brought very rapidly from a temperature T_1 to another temperature T_2. Depending upon whether $T_1 \lessgtr T_2$, either a supersaturation or an undersaturation of vacancies will result. Now, in this case, only the surface of the cylinder can serve as a sink or source for the vacancies. Therefore, the question to be answered is how quickly can the vacancies diffuse from the interior of the whisker to the surface in the case of cooling, or how quickly can they move from the surface to the interior in the case of heating. Fick's second law (see eq. (6-11)), when properly formulated, can be used to solve this cylindrical diffusion problem if it is assumed that (a) the vacancy diffusion coefficient D_V is independent of concentration, (b) because of the low concentration of vacancies in the metal they form an ideal dilute solution, and (c) the following initial and boundary conditions apply: 1. At time $t = 0$ following the change in temperature there exists a homogeneous distribution of the vacancies. 2. For times $t > 0$, vacancy equilibrium with $c_V = c_V^\circ$ is maintained at the surface $r = r_0$, and 3. $(\partial c_V/\partial r)_{r=0} = 0$ at all times at the axis of the cylinder $r = 0$. The explicit solution of this diffusion problem can again be found in the literature [5]. For times which are not too small, the relaxation time τ is given as:

$$\tau = \frac{r_0^2}{5.76\, D_V} \tag{7-2}$$

and so τ can be calculated if measurements of the vacancy diffusion coefficient D_V are available. The simplest way to measure this relaxation time is by recording the electrical resistance after the change in temperature, since all other contributions to the total change in resistance, except that arising from the change in concentration of vacancies, will occur instantaneously in the experiment as described. Furthermore, if the absolute contribution which a vacancy makes to the electrical resistance can be determined by means of quantum mechanical calculation [13], then the vacancy concentration can be determined, and so, by use of eq. (5-19), the self-diffusion coefficient of the metal can be calculated from the measured value of the vacancy diffusion coefficient.

Of course, the experiment which has just been described is only hypothetical. In practice, the outer surface of the metal serves as a vacancy source or sink only to an insignificant degree. This role is performed to a much greater extent by the dislocations and the low- and high-angle grain boundaries. We assume that the network of these sources and sinks remains fixed during the approach to equilibrium, and that relaxation processes do not cause the geometrical configuration of this network to change with time. If this is the case, then it is again possible, in principle, to calculate the kinetics of the equilibration of defects, provided that the spatial structure of the dislocation network is known. In order to obtain a useful first approximation, we introduce the parameter β which is the average radius of influence of a dislocation upon vacancies. This quantity was defined by eq. (6-10), and was discussed at greater length in section 6.1.2. Fick's second law in the form of eq. (6-11) can then be solved with the same initial and boundary conditions which were given with eq. (6-11). In this way, the following expression for the relaxation time for the equilibration of vacancies on the dislocation network in metals is obtained:

$$\tau = \frac{f(\varrho_D, \alpha)}{\pi \varrho_D D_V} \tag{7-3}$$

where ϱ_D is the dislocation density, α is the radius of the dislocation core, and $f(\varrho_D, \alpha)$ is a function of ϱ_D and α which follows from the rigorous and complete theory [14]. If a value of $\alpha = 10^{-7}$ cm is assumed, then $f(\varrho_D, \alpha)$ will vary only between 7 and 2.8 as the dislocation density varies from 10^1 to 10^8 per cm^2. The physical reason behind this, of course, is the fact that the main resistance to diffusion of the vacancies occurs in the immediate surroundings of the dislocation core. There is a narrow pass near the core, so to speak, through which all vacancies must squeeze their way. Among other things, this shows how little the approximation to the relaxation time depends upon an accurate estimate of the distribution of the dislocations in the network, and ultimately, this justifies the assumptions in the calculations. For normal dislocation densities in metals, which are of the order of 10^6 per cm^2, and for $D_V \approx \approx 10^{-6}$ cm^2/sec, which is a reasonable value at moderate temperatures, the relaxation time can be calculated by means of eq. (7-3) to be of the order of 1 sec.

The equilibration of interstitial atoms can be treated in exactly the same way. The existence of interstitial atoms is very improbable in the case of simple close-packed metals, however, since the defect energy of interstitial atoms in close-packed lattices is much higher than that of vacancies. However, if enough energy is supplied, as, for example, by means of plastic deformation or by irradiation of metals in a reactor, then Frenkel pairs (i.e. vacancies and an equal number of interstitial atoms) can be produced in metal crystals just as in ionic crystals. These Frenkel pairs are not in thermodynamic equilibrium. As long as their mobility is large enough, they will annihilate by means of a homogeneous reaction. The quantitative relationships given in section 6.1.1 also hold for this relaxation process. The same is true for equilibration occurring at grain boundaries which was treated quantitatively in section 6.1.2.

7.1.2. Interdiffusion in binary metal alloys and the Kirkendall effect

In order that we may deal with a concrete situation, let us consider the experimental arrangement shown in Fig. 7-1. Two metal cylinders A and B are welded flush against one another.

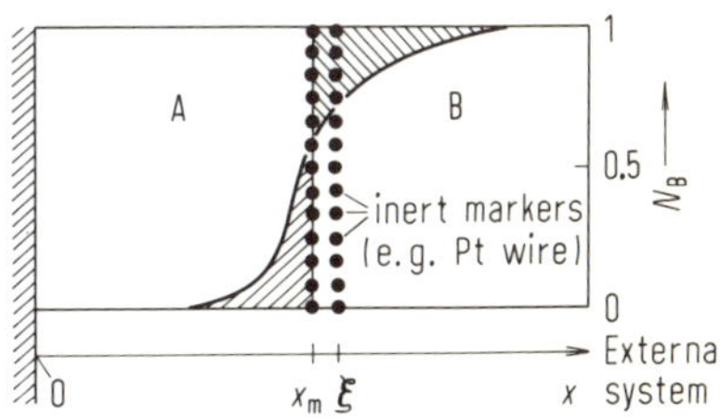

Fig. 7-1. Experimental configuration and coordinate system for the study of interdiffusion in simple metals; N_B = mole fraction of B, x_m = Matano interface, ξ = coordinate of the markers, or the marker plane.

A and B should form a complete range of solid solutions. This means that they should have the same crystal structure as well as similar molar volumes. The phenomenological transport problem here is concerned with the solution of Fick's laws for the given experimental conditions in this inhomogeneous system. The atomistic problem is concerned with the interpretation of the chemical diffusion coefficient which, for example, might have been determined by a Boltzmann-Matano analysis. It was shown in section 5.5.3 that, for the case of binary diffusion via vacancies, the chemical diffusion coefficient may be written as:

$$\tilde{D} = (N_2 D_1 + N_1 D_2)\left(1 + \frac{d \ln \gamma_1}{d \ln N_1}\right) \tag{7-4}$$

as long as the molar volume is constant and internal equilibrium is maintained. In order that eq. (7-4) be unconditionally applicable, it is necessary that all correlation effects (see section 5.4) can be neglected. Systematic deviations have been observed, especially for binary liquid phases. As can be seen, $\tilde{D}$ can be broken down into two terms. The latter of these depends only upon the thermodynamics of the binary system, and is thus called the "thermodynamic factor". There are two important special cases:

1. For the case of an ideal (Raoultian) solution or an ideal dilute (Henryan) solution, γ_1 is constant, and the thermodynamic factor is unity. Under these conditions:

$$\tilde{D} = (N_2 D_1 + N_1 D_2) \tag{7-5}$$

For the case of very dilute solutions with $N_2 \to 0$ and $N_1 \to 1$:

$$\tilde{D} \simeq D_2 \tag{7-6}$$

That is, in such solutions the chemical diffusion coefficient is equal to the component diffusion coefficient of the solute. This can easily be measured experimentally with the use of tracers. It should be noted that in the two preceding equations the coefficients D_i are, in general, concentration dependent. No simple solutions of Fick's second law can thus be expected in the systems just described.

2. For the case of regular solutions, the thermodynamic factor is $\left[1 + \frac{2\,\Omega}{RT} N_1(1-N_1)\right]$.

The assumption of regular solution behaviour is frequently a good approximation when the difference in the interaction energy between like and unlike metal atoms of a binary alloy is not too great. Ω is given by the quasichemical theory, and can be simply related to the pair formation energy between like and unlike atoms. Thermodynamically, Ω is simply the negative of the partial molar enthalpy of mixing of component 1 when $N_1 \to 0$ [6]. This value must be of the order of RT in order that an ideal entropy of mixing may be assumed as is required by the definition of a regular solution. Thus, once again, the thermodynamic factor is of the order of unity, and eq. (7-5) holds approximately. For all other cases, the thermodynamic factor can take on significant values, particularly for the case of phases which have narrow ranges of homogeneity. However, if the thermodynamics of the phases which are being formed in the diffusion experiment is known, then the thermodynamic factor can always be calculated. This leaves us with the term $(N_2 D_1 + N_1 D_2)$ which is concerned only with the transport properties of the system. The factors D_1 and D_2 in this term are a measure of the defect concentrations as functions of composition (in simple metals this usually means the vacancy concentration), and of the jump frequencies of atoms of type 1 and 2, also as functions of composition. We can thus appreciate how complicated the coefficient $\tilde{D}$ is. It is only possible to make definite predictions regarding the value of $\tilde{D}$ in certain limiting cases at best. For a dilute solution of component 2, for example, (when eq. (7-6) applies), the equation $D_2 = D_2^\circ \cdot (1 + \alpha N_2)$ can be used. That is, D_2 is a linear function of composition. Only in the zeroth approximation, however, can D_1 be taken as a linear function of the mole fraction N_2 even if N_2 is small [15, 16].

In order to obtain a deeper understanding of diffusion in inhomogeneous metallic systems in general, and of eq. (7-4) in particular, we must look more critically at the assumptions which we have been making. Most importantly, we must examine the reference system for the measurement of the particle fluxes, and we must look more closely at the assumption of local defect equilibrium. First of all, let us retain the assumption of a constant molar volume. Then, in Fig. 7-1, let us place the origin of the coordinate system $x = 0$ at the end of the cylinder A. This end lies far enough from the initial A/B boundary so that we can assume that B atoms never reach this point during the diffusion experiment. Since no atoms are ever lost during the experiment, it follows that the number of lattice sites and the dimensions of the sample remain constant if we neglect a possible very small change in the equilibrium concentration of defects which results from changes in concentration in the diffusion zone. This coordinate system can be called an "external system", since it is the system relative to an observer standing outside the diffusion zone. It is furthermore possible to mark the system of lattice sites through the use of inert markers. For example, very thin Pt or Mo wires can be inserted beforehand into the diffusion couple as shown in Fig. 7-1. In this way, the flux of particles relative to this system of lattice sites can be observed. In the year 1947, Smigelskas and Kirkendall [17] found that the system of lattice sites in the diffusion zone moved relative to the fixed external system for the diffusion couple A = Cu, B = brass. It was later shown that this phenomenon is completely general in metallic systems with a simple system of lattice sites. In systems, such as ionic systems, with more than one sublattice, the situation is less clear, inasmuch as it must first be established to which sublattice the inert markers belong. However, for the case of simple metal lattices, the "Kirkendall effect" says that, in a certain region of the diffusion zone, lattice sites are being created, and in another region they are being destroyed. This conclusion follows from the fact that the total number of lattice sites must remain constant. For the system Cu-brass or Cu-Zn, the explanation is as follows. Both Cu and Zn atoms are mobile and move via the same vacancies. However, for a given composition, the mobility of the Zn atoms is greater than that of the Cu atoms. This means that Zn atoms exchange sites with neighbouring vacancies faster then Cu atoms exchange sites with neighbouring vacancies. The result is an overall flux of vacancies in the direction towards the higher concentration of the faster partner. If local defect equilibrium is maintained, this means that the vacancies which arrive in the region of higher Zn concentration are being destroyed at sites of repeatable growth. As a result, the number of lattice sites in this region is decreasing, and the system of lattice points is contracting. The opposite process is occurring in the copper-rich side of the diffusion couple. Here, the vacancy concentration is less than its equilibrium value, and so vacancies are being formed on surfaces, dislocations, etc. In a more specific sense, then, the "Kirkendall effect" results from the maintenance of local defect equilibrium – a problem which, as we have seen, arises over and over again whenever we treat chemical reactions in the solid state quantitatively. Complications may arise when the deviations from defect equilibrium become great enough. For example, pores might form in the diffusion zone, projections could rise from the surface, or the sample might neck down.

The local lattice velocity v_L which arises as a result of the production and annihilation of vacancies is given by the following expression as shown in section 5.5.3 (see eq. (5-35)):

$$v_L = -(D_B - D_A)\, V_m \left[1 + \frac{d \ln \gamma_A}{d \ln c_A}\right] \frac{d c_A}{d x} \tag{7-7}$$

(V_m=molar volume). That is, v_L is proportional to the concentration gradient and to the difference between the component diffusion coefficients, as we might expect from the preceding discussion. From this it follows that if the metal atoms have equal mobilities or equal exchange frequencies with the vacancies at given alloy compositions, then no displacement of the markers in the external system will be observed. By using the expression $v_i = j_i / c_i$ (see eq. (5-32)), and by noting that the lattice velocity v_L is caused by a local vacancy flux, we can immediately calculate this vacancy flux j_V as:

$$j_V = \frac{v_L}{V_m} = -(D_B - D_A)\left[1 + \frac{\mathrm{d}\ln\gamma_A}{\mathrm{d}\ln c_A}\right]\frac{\mathrm{d}c_A}{\mathrm{d}x} \tag{7-8}$$

According to eq. (7-7), we can measure the difference between the two diffusion coefficients $D_B - D_A$ in a simple experiment by observing the displacement of inert markers which were originally situated in the plane of the weld at the beginning of the diffusion experiment (see Fig. 7-1). According to eq. (5-53), the geometrical position of this plane in the external system is at all times coincident with the Matano interface. Therefore, we are concerned with measuring the motion of the markers relative to the Matano plane. The markers are fixed in the lattice system. Therefore they denote a plane of constant concentration. For the experimental arrangement given in Fig. 7-1, $c(y) \equiv c(x/\sqrt{t})$ is therefore constant. That is, the marker plane, which has the coordinate ξ in the external system, moves proportionately to $\sqrt{t}$ away from the Matano plane x_m into the region of the higher diffusion coefficient D_i. If $c(y)$ is constant, then $(\mathrm{d}c/\mathrm{d}y)_y$ is also constant, and can be considered to be known when a solution $c = c(y)$ is known. Accordingly, by setting $\left(\frac{\partial c}{\partial x}\right) = \frac{1}{\sqrt{t}}\left(\frac{\partial c}{\partial y}\right) \propto \frac{1}{\sqrt{t}}$ into eq. (7-7), we obtain the result that $v_L(\xi)$ is proportional to $1/\sqrt{t}$, and, except for the factor $(D_B - D_A)$, the constant of proportionality contains only quantities which are known.

If the concentration gradients are very steep, and if the density of sources and sinks for vacancies is relatively small, then spontaneous formation of pores can occur in the metal as a result of a local supersaturation of vacancies. The conditions for this situation to occur have been examined in some detail [18]. As long as the first two factors in eq. (7-8) vary monotonically, then the production and annihilation of vacancies $\mathrm{d}j_V/\mathrm{d}x$ depend essentially only upon the curvature of the concentration profile $\mathrm{d}^2 c_A/\mathrm{d}x^2$, as follows from the differentiation of eq. (7-8). This is demonstrated in Fig. 7-2.

The "Kirkendall effect" should be understood not just as a displacement of markers in the common experimental arrangement in Fig. 7-1, but rather as the ability of the lattice to respond to deviations from the local equilibrium concentration of vacancies by the production or annihilation of vacancies (and of lattice sites). Thus, the proper quantitative measure of the "Kirkendall effect" is $\mathrm{d}j_V/\mathrm{d}x$.

In general, the existence of a Kirkendall effect in the sense of a displacement of markers does not permit a conclusion to be drawn as to a definite diffusion mechanism as, for example, a vacancy diffusion mechanism. To give a very improbable example – even if a ring mechanism is the predominant diffusion mechanism, a displacement of markers in the external system could nevertheless occur because lattice sites can be locally created or destroyed at sites of repeatable growth as a result of the fact that the concentration of vacancies, which are always present, depends upon the composition of the alloy.

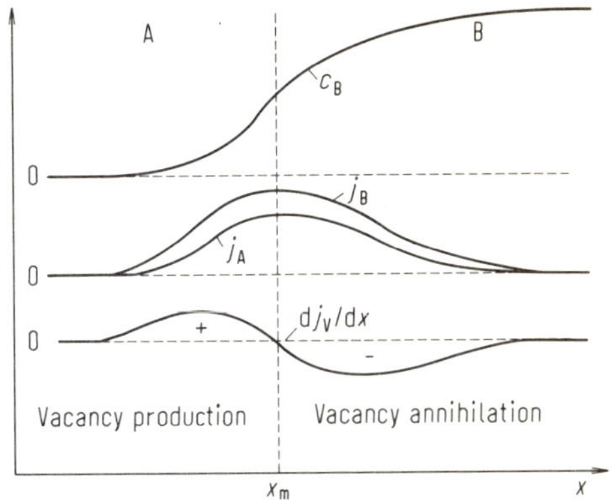

Fig. 7-2. The concentration profile (c_B), the particle fluxes (j_A, j_B), and the production or annihilation of defects (dj_V/dx) in the diffusion couple A-B (see Fig. 7.1).

Reference should be made here to certain observations [18] which show that the assumption of local defect equilibrium does not always apply. In certain metallic systems (e.g. Ag-Au), the interaction between dislocations and vacancies is so small that the lifetime of the vacancies is long enough to permit them to diffuse out of the actual diffusion zone. This would give rise to a displacement of markers even if they had been placed outside the diffusion zone shown in Fig. 7-2 before the diffusion experiment was started, and so the assumptions upon which eqs. (7-7) and (7-8) were based would no longer be applicable.

Finally, to conclude this section, let us discard the assumption that the molar volume V_m is constant during interdiffusion. The definition of the chemical diffusion coefficient $\tilde{D}$ in eq. (5-33) does not require such an assumption, although this assumption was used in the derivation of the special form of Darken's equation (5-37) or (7-4). What procedure can be used to calculate the chemical diffusion coefficient from experimental diffusion profiles, for example along the lines of a Boltzmann-Matano analysis, when V_m varies with composition? The successful solution of this problem has been reported in a few recent articles [19, 20, 21]. The application of the formulae in these articles is especially convenient, since no explicit calculation of the position of the Matano interface is necessary. Once again, the initial configuration is a diffusion couple with two semi-infinite regions, and a discontinuity in concentration at time $t = 0$ (cf. section 5.5). The components here are designated 1 and 2 rather than A and B. The definition of the coordinate of the Matano plane x_m in systems with variable molar volumes relative to an arbitrary origin of coordinates is given by rearrangement of eq. (5-53) as:

$$\int_{-\infty}^{x_m} \frac{N_2 - N_2^-}{V_m} \, dx + \int_{x_m}^{+\infty} \frac{N_2^+ - N_2}{V_m} \, dx = 0 \tag{7-9}$$

N_2^- is the mole fraction of component 2 for $x = -\infty$, and N_2^+ is the mole fraction for $x = +\infty$. If the auxiliary variable Y is defined as:

$$Y = \frac{N_2 - N_2^-}{N_2^+ - N_2^-} \tag{7-10}$$

then the mole fractions N_1 and N_2 can also be written as:

$$N_1 = (1 - N_2^+)\, Y + (1 - N_2^-)\,(1 - Y)$$
$$N_2 = N_2^+\, Y + N_2^-\,(1 - Y) \tag{7-11}$$

For the subsequent calculations, reference should be made to the original literature. The resultant expression for the concentration dependent diffusion coefficient $\tilde{D}(N_2)$ at the space coordinate $x(N_2)$ with molar volume $V_m(N_2)$ and auxiliary variable $Y(N_2)$ is:

$$\tilde{D}\,(N_2) = \frac{(N_2^+ - N_2^-)\, V_m\,(N_2)}{2\,t\,(\partial N_2/\partial x)_{x(N_2)}} \cdot \left[(1 - Y\,(N_2)) \cdot \int_{-\infty}^{x(N_2)} \frac{Y}{V_m}\,\mathrm{d}x + Y\,(N_2) \cdot \int_{x(N_2)}^{+\infty} \frac{1 - Y}{V_m}\,\mathrm{d}x\right] \tag{7-12}$$

Only the differential $\mathrm{d}x$ appears in eq. (7-12). Therefore, as stated above, $\tilde{D}(N_2)$ can be determined in a coordinate system with an arbitrary origin, and it is not necessary to determine the position of the Matano interface. If $V_m\,(N_2)$ is known as a function of the mole fraction, and if the diffusion profile $N_2\,(x)$ has been measured, then eq. (7-12) can be used to determine the chemical diffusion coefficient $\tilde{D}(N_2)$. This could be done graphically, for instance, by a modification of the procedure illustrated in Fig. 5-8.

7.1.3. Diffusion in inhomogeneous ternary systems

This is an area of study of very great practical importance, since all technologically interesting metal alloys consist of three or more components. The phenomenological equations necessary to describe the diffusional processes in this case have already been presented in section 5.2. From the thermodynamics of irreversible processes it may be shown that, in a suitably chosen coordinate system, the diffusion fluxes are linear functions of the chemical potential gradients of all components. The transport coefficients L_{ik} satisfy the so-called Onsager relationships: $L_{ik} = L_{ki}$. Proceeding once again from the general system of equations (5-2) to a set of practical equations which are written explicitly in terms of concentrations (analogously to Fick's laws for the case of binary diffusion), we obtain a system of second-order non-linear differential equations which apparently bear a perfect resemblance to the simple form of Fick's second law. For an isothermal, isobaric, one-dimensional diffusion problem, this set of equations is as follows:

$$\frac{\partial c_1}{\partial t} = \frac{\partial}{\partial x}\left[\tilde{D}_{11}\frac{\partial c_1}{\partial x} + \tilde{D}_{12}\frac{\partial c_2}{\partial x} + \ldots\right]$$
$$\frac{\partial c_2}{\partial t} = \frac{\partial}{\partial x}\left[\tilde{D}_{21}\frac{\partial c_1}{\partial x} + \tilde{D}_{22}\frac{\partial c_2}{\partial x} + \ldots\right] \text{ etc.} \tag{7-13}$$

For a reaction in a ternary single-phase inhomogeneous metallic system with given initial and boundary conditions, it may easily be appreciated that finding a mathematical solution of eq. (7-13) in the form $c_i = f(\tilde{D}_{ik}, x, t)$ is incomparably more difficult than in the case of binary alloys. If the starting substances only differ slightly in composition, then constant values of $\tilde{D}_{ik}$ equal to the average values over the small concentration interval can be introduced into eq. (7-13) with the resultant simplification:

$$\frac{\partial c_1}{\partial t} = \tilde{D}_{11} \frac{\partial^2 c_1}{\partial x^2} + \tilde{D}_{12} \frac{\partial^2 c_2}{\partial x^2} + \ldots$$

$$\frac{\partial c_2}{\partial t} = \tilde{D}_{21} \frac{\partial^2 c_1}{\partial x^2} + \tilde{D}_{22} \frac{\partial^2 c_2}{\partial x^2} + \ldots \quad \text{etc.} \tag{7-14}$$

The generalized chemical diffusion coefficients $\tilde{D}_{ik}$ which have been introduced here are related to the transport coefficients in the general system of equations (5-2) as follows:

$$\tilde{D}_{ik} = \sum_j L_{ij} \frac{\partial \mu_j}{\partial c_k} \tag{7-15}$$

In the following discussion, it will be assumed once again, for the sake of simplicity, that no change in volume occurs during the interdiffusion. This appears to be approximately true for many metallic systems when they exhibit a broad range of miscibility. Then, depending upon the structure of the alloy, certain restrictions upon the fluxes in the external system can be stated. These restrictions greatly simplify the reaction problem in certain limiting cases. For a substitutional alloy, for instance, the following restriction on the fluxes of components 1, 2, and 3 may be written:

$$j_1 + j_2 + j_3 = 0 \tag{7-16}$$

For an interstitial alloy with component 1 on the interstitial sites:

$$j_2 + j_3 = 0 \tag{7-17}$$

Finally, for an interstitial alloy with both components 1 and 2 on interstitial sites:

$$j_3 = 0 \tag{7-18}$$

In order now to apply the preceding general considerations to a concrete but still relatively simple case, we shall discuss the Fe-Si-C system. We find here a situation in which a component can diffuse locally against its concentration gradient. This is known as "up-hill diffusion". Austenitic Fe-Si-C consists of a face-centered-cubic iron lattice with carbon on the interstitial sites. The silicon atoms are substituted on iron sites, and so the mobility of the silicon atoms is orders of magnitude smaller than that of the carbon atoms [22]. In Fig. 7-3 are shown the results of an experiment in which two iron cylinders with about the same carbon contents but with very different silicon contents were welded flush against one another and held for 13 days at 1050 °C. The experimental arrangement as well as the carbon concentration (N_C), the carbon activity (a_C), and the silicon concentration (N_{Si}) at the end of the experiment are shown.

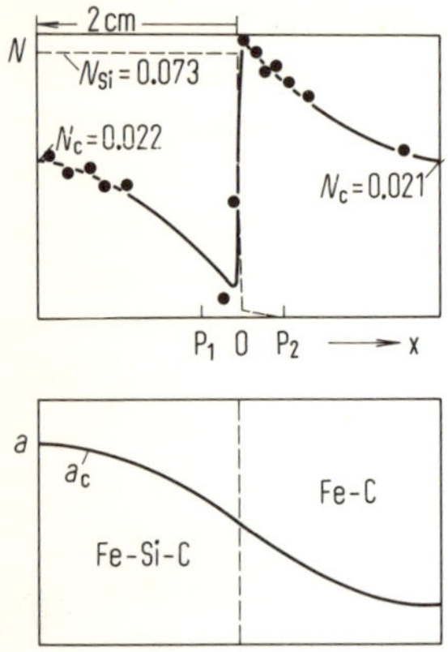

Fig. 7-3. Experimental arrangement, concentration profile, and schematic activity profile for an interdiffusion experiment in the system Fe-Si-C after Darken [22]. Experiment was conducted for 13 days at 1050 °C. 0 = weld plane; P_1, P_2 = arbitrary points.

The observed concentration profiles can be theoretically explained on the basis of the relationships (7-13) and (7-15) [23]. First of all we note that the limiting case of eq. (7-17) applies here. That is, silicon diffuses via vacancies, and carbon diffuses in the interstices of the fcc lattice. The partial differential equations for this case are as follows:

$$\frac{\partial c_{Si}}{\partial t} = \frac{\partial}{\partial x} \tilde{D}_{SiSi} \frac{\partial c_{Si}}{\partial x} + \frac{\partial}{\partial x} \tilde{D}_{SiC} \frac{\partial c_C}{\partial x} \tag{7-13a}$$

$$\frac{\partial c_C}{\partial t} = \frac{\partial}{\partial x} \tilde{D}_{CSi} \frac{\partial c_{Si}}{\partial x} + \frac{\partial}{\partial x} \tilde{D}_{CC} \frac{\partial c_C}{\partial x} \tag{7-13b}$$

By means of the substitution $y = x/\sqrt{t}$, these equations can be converted into ordinary differential equations. An analogous equation could also be written for the change in concentration of iron. However, because of eq. (7-17) and because of thermodynamic relationships in the ternary system (Gibbs-Duhem equation), this equation would not be independent. Because of the higher mobility of carbon, it is observed that the carbon reaches equilibrium after a relatively short time (as compared to the diffusion of silicon). From eq. (7-15) we may write:

$$\tilde{D}_{SiC} = L_{SiSi}\left(\frac{\partial \mu_{Si}}{\partial c_C}\right) + L_{SiC}\left(\frac{\partial \mu_C}{\partial c_C}\right) \tag{7-15a}$$

In a dilute solution of Si and C in Fe, it is certainly true that $L_{SiC} \ll L_{SiSi}$ and that $(\partial \mu_{Si}/\partial c_C) <$ $< (\partial \mu_C/\partial c_C)$. Thus, in the first approximation, $\tilde{D}_{SiC}$ can be set equal to zero so that eq. (7-13a) can be written as the following ordinary differential equation:

$$\frac{\mathrm{d} c_{Si}}{\mathrm{d} y} = -\frac{2}{y} \frac{\mathrm{d}}{\mathrm{d} y} \tilde{D}_{SiSi} \frac{\mathrm{d} c_{Si}}{\mathrm{d} y} \tag{7-19}$$

The solution of this equation under the assumption that $\tilde{D}_{SiSi}$ is constant is as follows (see eq. (5-42)):

$$c_{Si} = \frac{c_{Si}^{\circ}}{2}\left[1 - \operatorname{erf}\frac{x}{2\sqrt{\tilde{D}_{SiSi}\, t}}\right] \tag{7-20}$$

In other words, this means that in this special case the silicon atoms move as if they were unaware of the carbon atoms and were concerned only with equalizing their own concentration. On the other hand, a rigorous solution of the equations for the case of carbon shows that the carbon concentration is given as a function of $y = x/\sqrt{t}$ by a relatively complicated expression in which the difference of two error functions erf $(x/2\sqrt{D_{ii}t})$ occurs. However, the theoretical analysis is confirmed by experiment as can be seen from the calculated curve which is shown together with the experimental points in Fig. 7-3 [23]. The physico-chemical explanation for the entire phenomenon is based on the fact that the carbon activity a_C is higher in the alloy containing silicon than in the alloy without silicon for the same carbon content. The effect of a small concentration of a third element (Si) upon the activity of a dissolved substance (C) can be described by means of the so-called interaction parameter:

$$\varepsilon_{C(Fe)}^{Si} = \left(\frac{\partial \ln \gamma_C}{\partial N_{Si}} \right)_{a_C}$$

γ_C is the activity coefficient.

In the present case, this parameter is positive. This means that the carbon content must decrease in the silicon-rich region and must increase in the silicon-poor region of the iron alloy until the carbon activity is everywhere uniform. At the welded junction, the local equilibration of carbon activity is achieved practically instantaneously. The shape of the concentration profiles of carbon in Fig. 7-3 then can also explain in a non-rigorous way why the mathematical solution for the carbon concentration profile is in the form of two error functions. This phenomenon of multicomponent diffusion, which is very complicated even in the present case of diffusion in a system with three components, can only be completely understood by a study of the original references [4, 22, 23, 34].

In order to further illustrate the diffusional process in a single-phase ternary system, use can be made of the Gibbs triangle. The path taken by the composition of the alloy as a function of time at a certain point on the experimental sample can be drawn on the triangle [23]. In the experiment discussed above, we might consider two points which lie to the right and to the left of the welded junction. These points, P_1 and P_2, and the original concentrations at these points, are shown in Fig. 7-3. At time $t \rightarrow \infty$, the composition of the now homogeneous couple will lie at a point in the Gibbs triangle on a straight line joining the initial compositions of the two parts of the couple. The exact position of the point in the triangle will depend upon the volumes of the two parts. In the special case of Fe-Si-C, the concentrations at the two points will first of all move along lines of constant silicon content. This is a direct result of the relatively low mobility of the silicon atoms. In a similar way, the variation of composition with position at a fixed time can also be represented on the Gibbs triangle. These curves must cross the line joining the initial concentrations of the diffusion couple at least once. In single-phase regions, S-shaped curves are always found if the $\tilde{D}_{ik}$ are sufficiently small relative to the $\tilde{D}_{ii}$. This is illustrated in Fig. 7-4 for the case of the diffusion experiment which was shown in Fig. 7-3.

In practice, an analysis of diffusional processes in multiphase multicomponent systems might almost be considered to be more important than the description of the processes in single-phase systems. Even if the initial materials A and B of the diffusion couple are located in Fig. 7-4 in a one-phase field, a reaction path in the Gibbs triangle, drawn just as in Fig. 7-4, can still cross over multiphase regions. It is only very recently that a beginning has been made

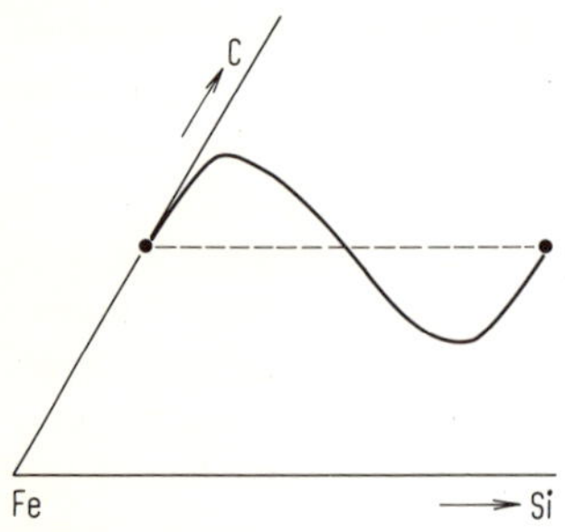

Fig. 7-4. Schematic representation of the composition as a function of position for the diffusion experiment shown in Fig. 7-3.

towards the quantitative description of such problems [23]. The following qualitative comments may give some insight into the complexity of these problems: For the relatively simple case of alloy oxidation, $(A,B)+1/2\,O_2=(A,B)O$, it is known that planar interfaces between the metallic phase (A, B) and the oxide are not always stable during growth. Depending upon the transport coefficients and the thermodynamics, the interfaces can become increasingly uneven as the reaction proceeds. That is, they become unstable [24]. Since this system is a special case of a ternary system, it is immediately obvious that these observations can be generalized: If the reaction path in the Gibbs triangle crosses the boundary of a single-phase region, then planar interfaces are not stable in every case. Furthermore, if we bear in mind that barriers to nucleation can lead to supersaturation, we can easily appreciate that very complicated morphological structures can arise in an initially one-phase ternary diffusion couple when the reaction path crosses multiphase regions. Examples can be found in two recent works on ternary diffusion in the Fe-Ni-Cr [35] and the Cu-Zn-Ni [36] system which illustrate the above remarks.

7.2. Diffusion in multiphase alloy systems

In this section we shall once again be concerned with binary metallic systems. In the previous section, some general remarks were made regarding diffusion in multiphase multicomponent systems. The complexity of the problem was pointed out, and the difficulties entailed in a quan-

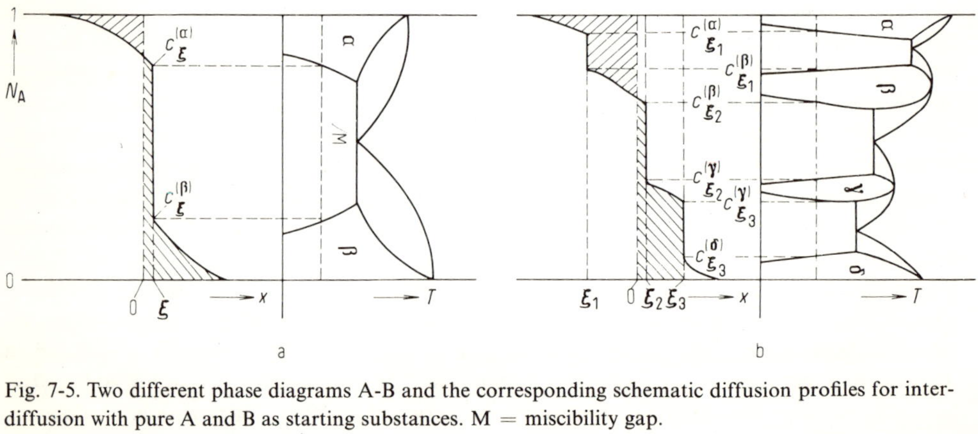

Fig. 7-5. Two different phase diagrams A-B and the corresponding schematic diffusion profiles for interdiffusion with pure A and B as starting substances. M = miscibility gap.

titative treatment were discussed. Since our present purpose is to obtain quantitative expressions for solid state reactions, however, we shall limit ourselves in this section to a discussion of binary systems. Two important limiting cases can be distinguished. These are illustrated in Fig. 7-5a and 7-5b.

In the first case (a), a miscibility gap M occurs in the phase diagram, and there will thus be a phase boundary in the diffusion couple with a discontinuity in concentration $c_\xi^{(\alpha)} - c_\xi^{(\beta)}$. In the second case (b), there are several phases in the phase diagram, and several phase boundaries with discontinuities in concentration occur in the diffusion couple [25]. In the limiting case, the phases in Fig. 7-5a and 7-5b may have only very narrow ranges of homogeneity. The theoretical expressions will thus be simplified, inasmuch as the end phases α and β can then be treated as ideal dilute solutions, and the rest of the phases can be treated as nearly stoichiometric compounds. All the cases just mentioned will now be discussed in detail.

7.2.1. Diffusion in a binary two-phase system

The simplest case occurs when two substances which are only very slightly soluble in one another, such as Al-Ge [7], are pressed together and annealed at diffusion temperatures (400 °C for Al-Ge). The chemical diffusion coefficient $\tilde{D}$ is then equal to the diffusion coefficient of the solute metal, which may be assumed to be independent of concentration, and which can be directly measured, for example by means of radioactive tracers or stable isotopes. A radioactive tracer atom moves in the dilute solution on the average in exactly the same way as a solute atom moves in the solvent. It thus follows from eq. (7-4) that:

$$\tilde{D} = N_2 D_1 + N_1 D_2 \simeq D_2 = D_2^* \qquad (7\text{-}21)$$

We have neglected here an effect which arises from the fact that even in a dilute solution in the diffusion couple in an interdiffusion experiment there is a chemical concentration gradient, whereas in a tracer experiment no such concentration gradient is present, and all that takes place is an exchange of active and inactive particles. This effect is of low-order, however, and disappears in the limit $N_2 \to 0$.

The diffusion coefficient D_1 of the solvent can be determined from the tracer coefficient D_1^* by the use of the correlation factor f and the relationship $D_1 f = D_1^*$ as in the case of the pure substance. However, D_1 is not required in the description of the chemical reaction (i.e. in Fick's equations) for this case of dilute solutions. Methods exist whereby the changes in the values of D_1 and D_2 on going from dilute to no longer dilute metallic solutions with the addition of solute 2 can be estimated. These approximations are based upon detailed atomic models of the same sort as are used to calculate jump frequencies of individual atoms or defects from measured tracer diffusion coefficients. In the first approximation, the diffusion coefficients are given by expressions of the form $D_i \simeq D_i^\circ (1 + \alpha_i N_2)$ [15, 16], where α is a factor of proportionality. This type of expression has already been discussed.

For the two-phase system presently being considered, under the conditions as stated, the chemical diffusion coefficient is nearly constant, and is given by eq. (7-21). For sufficiently long samples, then, the solution of the diffusion equations can be adapted from eq. (5-42), and so we can leave this simplest case.

The next type of reaction between metals to be discussed in a systematic treatment is the reaction between a pure metal A and a two-phase mixture whose composition lies in the miscibility gap between the α and β solutions. The quantitative treatment of this case, however, is a good deal more difficult than for the preceding case. The main question of practical interest is how fast does the boundary between the single-phase α region and the two-phase $(A, B)^{(\alpha)} + (A, B)^{(\beta)}$ region (i.e. the reaction front) advance. In Fig. 7-6 are shown the details of the experimental arrangement, the position of the coordinate system, and the scheme of notation.

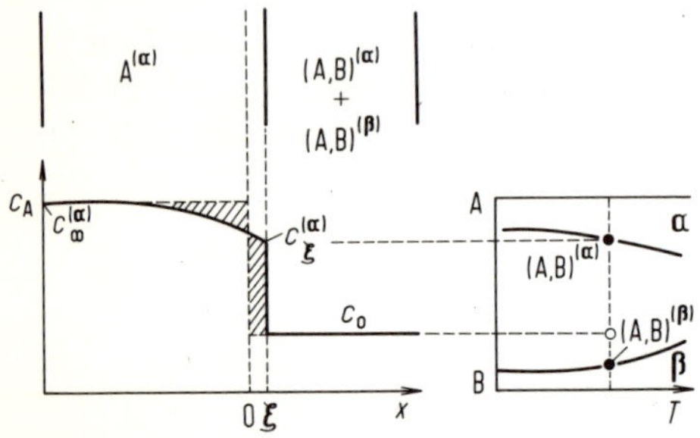

Fig. 7-6. The experimental arrangement and the concentration profile for the reaction between A and a nonvariant two-phase mixture $(A, B)^{(\alpha)}$ and $(A, B)^{(\beta)}$.

The following simplifying assumptions are also made: the chemical diffusion coefficient in the α-phase is constant; local thermodynamic equilibrium is maintained at the phase boundaries (during the early stages of the reaction this will only hold approximately because of the high concentration gradients); and the volumes of mixing are negligible. The last assumption holds true for ideal dilute solutions, and in many metallic systems it is a sufficiently good approximation. Fick's second law (eq. (5-30)) applies in the single-phase α region. The condition of continuity also applies:

$$(c_\xi^{(\alpha)} - c_0)\frac{\mathrm{d}\xi}{\mathrm{d}t} = -\tilde{D}^{(\alpha)}\left(\frac{\partial c}{\partial x}\right)_\xi^{(\alpha)} \tag{7-22}$$

This equation says that, at the reaction front, the entire flux of A crossing the boundary from the α-phase, $j_A = -\tilde{D}^{(\alpha)}(\partial c/\partial x)_\xi^{(\alpha)}$, is used up in transforming the phase $(A, B)^{(\beta)}$ into $(A, B)^{(\alpha)}$, that is, in transforming $(A, B)^{(\beta)}$ into the α-phase with composition $c_\xi^{(\alpha)}$. In a two-phase region in a binary system there are no further degrees of freedom once the pressure P and reaction temperature T are fixed. Therefore, no diffusion can take place in this region. It can easily be proven by substitution that both Fick's second law and the condition of continuity (7-22) are satisfied by the expression:

$$\xi = 2p\,(\tilde{D}^{(\alpha)} t)^{1/2} \tag{7-23}$$

which gives the position of the reaction front ξ as a function of time. p is a dimensionless parameter which can be calculated from eq. (7-22) and from the solution of Fick's second law in the α-region. According to eq. (5-42), this solution must be of the form:

$$c^{(\alpha)} = c_\infty^{(\alpha)} - f(\tilde{D}^{(\alpha)}, c_\xi^{(\alpha)})\,\mathrm{erf}\frac{x}{2\,(\tilde{D}^{(\alpha)} t)^{1/2}} \tag{7-24}$$

provided that the sample A is sufficiently long. $f(\tilde{D}^{(\alpha)}, c_\xi^{(\alpha)})$ is a constant which comes out of the calculations and which is dependent only upon the known quantities $\tilde{D}^{(\alpha)}$ and $c_\xi^{(\alpha)}$. An example of an explicit calculation will be given below.

So far we have formulated the reaction in an entirely general way. With only a slight modification of the boundary conditions, these general considerations can be applied to the important practical case in which two-phase alloys of ferrite and austenite are carburized or decarburized to give single-phase austenite or ferrite, as can be easily seen by reference to the iron-carbon phase diagram [7]. In Fig. 7-7, the experimental arrangement as well as the corresponding carbon concentration profile are shown.

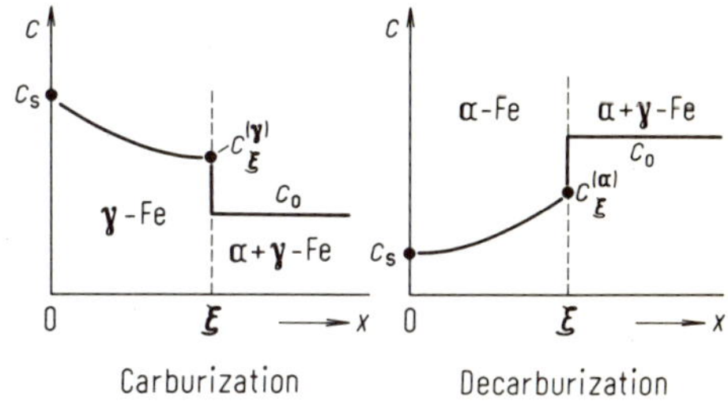

Fig. 7-7. The carburization or decarburization of iron-carbon alloys.

The surface concentration of carbon, c_S, is constant. This is the only boundary condition which is different from the previously discussed general case (Fig. 7-6). This surface concentration can be fixed by means of a CO/CO_2 or a CH_4/H_2 gas mixture. Fick's second law holds in the single-phase region, and eqs. (7-22) to (7-24) are equally applicable here as in the case of the reaction between A and the two-phase mixture $(A, B)^{(\alpha)} + (A, B)^{(\beta)}$. By substituting eqs. (7-23) and (7-24) into eq. (7-22) and using Fick's second law, with the given boundary conditions for the carburization or decarburization experiment, we obtain an equation for the unknown parameter p. For the case of carburization, this equation is:

$$\frac{c_s - c_\xi^{(\gamma)}}{c_\xi^{(\gamma)} - c_0} = \pi^{1/2} p e^{p^2} \operatorname{erf} p \tag{7-25}$$

For decarburization, $c^{(\gamma)}$ in eq. (7-25) is replaced by $c^{(\alpha)}$. p can be determined graphically from eq. (7-25) for any particular experiment. It bears repeating once again that the coordinate ξ of the reaction front moves according to a parabolic rate law, and that the diffusion profile in the single-phase region can be described by an error function erf $x/2\sqrt{\tilde{D}t}$ just as in the solution of Fick's second law for a semi-infinite region. The slight concentration dependence of the diffusion coefficient of carbon has been neglected [37], and the chemical diffusion coefficient has been taken to be constant.

The final case to be treated in this section is the reaction between two metals A and B whose phase diagram exhibits a miscibility gap and extended regions of solubility α and β as shown in Fig. 7-5. Once again, many practically important applications can be found. Examples are the systems Ag-Cu (800 °C), Au-Co (900 °C), Zn-Cd (250 °C), or Co-Cu (1000 °C). Consider two sufficiently long samples A and B which are welded together at the point $x = 0$. Assume that local equilibrium is achieved everywhere and at all times. After a diffusion time t there will be a discontinuity in concentration $c_\xi^{(\alpha)} - c_\xi^{(\beta)}$ at $x = \xi$, which can be calculated from the phase diagram. By applying the expression for $\tilde{D}$ which was obtained from the Boltzmann-Matano analysis (cf. eq. (5-52)) to both the α- and the β-phase at $x = \xi$, and then subtracting the two equations, we obtain, after slight rearrangement, the condition of continuity at the boundary $x = \xi$:

$$\tilde{D}^{(\alpha)} \left(\frac{\partial c}{\partial x}\right)_{\xi}^{(\alpha)} - \tilde{D}^{(\beta)} \left(\frac{\partial c}{\partial x}\right)_{\xi}^{(\beta)} = -\frac{\xi}{2t} \int_{c^{(\beta)}}^{c^{(\alpha)}} \mathrm{d}c = -\frac{\xi}{2t} (c_{\xi}^{(\alpha)} - c_{\xi}^{(\beta)}) \qquad (7\text{-}26)$$

Furthermore, Fick's second law applies in both the α- and β-phase with the boundary conditions for a semi-infinite region. Eq. (7-26) states that the volume of β-phase which is converted to α-phase per unit time is determined by the flux $j_A = -\tilde{D}^{(\alpha)}(\partial c/\partial x)_{\xi}^{(\alpha)}$ at the phase boundary, less the flux $j_A = -\tilde{D}^{(\beta)}(\partial c/\partial x)_{\xi}^{(\beta)}$ which is simultaneously diffusing away from the boundary into the β-phase. The solution of Fick's second law in both the α- and β-phases is again of the form of equation (7-24). Once again, the expression $\xi = 2p\,(\tilde{D}^{(\alpha)} t)^{1/2}$ satisfies all the relevant equations for this kinetic problem. With eq. (7-26) as the continuity condition in the present case, we obtain a complicated transcendental equation for $p = f(c_{\xi}^{(\alpha)}, c_{\xi}^{(\beta)}, \tilde{D}^{(\alpha)}/\tilde{D}^{(\beta)})$ in analogy with eq. (7-25). This equation can also be solved graphically for any particular experiment.

From the expression (7-23) for ξ, the rate of advancement of the phase boundary can be calculated as $\mathrm{d}\xi/\mathrm{d}t = v_{\xi} = p\,(\tilde{D}^{(\alpha)}/t)^{1/2}$. Through elimination of p it follows that $v_{\xi} = \xi/2t$. The meaning of the right-hand-side of eq. (7-26) now becomes clear. Finally, it may be noted from eq. (7-26) that $\mathrm{d}\xi/\mathrm{d}t$ becomes equal to zero when the diffusion currents j_A in the α- and β-phases at the phase boundary $x = \xi$ are equal at all times. The Matano interface is then coincident with the phase boundary. In all other cases, the phase boundary moves away from the Matano plane with velocity $v_{\xi} = \xi/2t$ as has just been calculated.

7.2.2. Formation of intermetallic phases with narrow ranges of homogeneity

In this section we shall consider first of all reactions of the type $n\mathrm{A} + m\mathrm{B} = \mathrm{A}_n\mathrm{B}_m$ in which the reaction product $\mathrm{A}_n\mathrm{B}_m$ has a narrow range of homogeneity. An example would be the reaction Al + Sb = AlSb at 350 °C. In general, we shall not be concerned with determining the chemical diffusion coefficient by a Boltzmann-Matano analysis, since in the limiting case the diffusion profile in the diffusion couple consists of a pure step function. Reactions of this type parallel reactions between ionic crystals as discussed in section 6.2. Two situations must be separately discussed:

1. The metal atoms A and B of the reaction product $\mathrm{A}_n\mathrm{B}_m$ occupy different sublattices. In metals there is no electrical coupling of the diffusion currents of the atomic components because there is no condition of electroneutrality to be obeyed as there is in ionic crystals. Thus, the divergence-free fluxes j_A and j_B in the reaction product are simply given relative to the lattice system by eq. (5-28). The component diffusion coefficients can, as usual, be determined from tracer diffusion measurements if correlation effects are accounted for. One should not forget that for intermetallic compounds the component diffusion coefficient D_i will depend to a greater or lesser extent upon the component activities. However, in the first approximation, the average value over the reaction layer can be used, especially when the component activities a_A and a_B within the region of homogeneity of a phase $\mathrm{A}_n\mathrm{B}_m$ do not vary by more than a power of ten. However, for the case of intermetallic compounds with very high Gibbs free energies of formation of $\mathrm{A}_n\mathrm{B}_m$ (e.g. NiAl), this simplification is hardly applicable. In this case, the

dependence of the component diffusion coefficient upon activity must be explicitly taken into account in the calculations. This will be given by the dependence of the defect concentration upon activity as in eq. (6-24). However, if average values can be used, then the growth law for the reaction product can be calculated from eq. (5-28) and from the fact that the rate of growth of the product layer is related to the fluxes in the reaction product by the equation:

$$\frac{\mathrm{d}\Delta x}{\mathrm{d}t} = \left(\frac{1}{n} j_\mathrm{A} + \frac{1}{m} j_\mathrm{B}\right) V_\mathrm{m} \tag{7-27}$$

The resultant rate equation for the increase in thickness of the reaction product layer in a one-dimensional case is:

$$\frac{\mathrm{d}\Delta x}{\mathrm{d}t} = -\left(\frac{\bar{D}_\mathrm{A}}{n} + \frac{\bar{D}_\mathrm{B}}{m}\right) \frac{\Delta G^0_{\mathrm{A}_n\mathrm{B}_m}}{RT} \frac{1}{\Delta x} \tag{7-28}$$

The bars over the component diffusion coefficients indicate that these are average values. In the derivation of eq. (7-28) the following formula was used:

$$\Delta \mu_\mathrm{A} = \frac{1}{n} \Delta G^0_{\mathrm{A}_n\mathrm{B}_m} \tag{7-29}$$

$\Delta \mu_\mathrm{A}$ is the difference between the chemical potentials of A at the two phase boundaries of the reaction product, and $\Delta G^0_{\mathrm{A}_n\mathrm{B}_m}$ is the standard free energy of formation of $\mathrm{A}_n\mathrm{B}_m$ from the elements. This formula applies as long as local thermodynamic equilibrium is maintained, and it can easily be derived from the equilibrium condition for the reaction $n\mathrm{A} + mB = \mathrm{A}_n\mathrm{B}_m$. Integration of eq. (7-28) yields the parabolic reaction rate law in the form $\Delta x^2 = 2\bar{k}t$, and the parabolic (practical) reaction rate constant is given by:

$$\bar{k} = \left(\frac{\bar{D}_\mathrm{A}}{n} + \frac{\bar{D}_\mathrm{B}}{m}\right) \frac{\Delta G^0_{\mathrm{A}_n\mathrm{B}_m}}{RT} \tag{7-30}$$

It is frequently found that $D_\mathrm{A} \gg D_\mathrm{B}$ (or $D_\mathrm{A} \ll D_\mathrm{B}$), in which case the reaction layer grows only at the phase boundary with B (or A), and so eq. (7-30) can be simplified. Parabolic growth of intermetallic compound layers was observed long ago in the systems Au-Sn, Cd-Cu, and Fe-Zn by Tammann [38].

If marker experiments are performed for the case in which $D_\mathrm{A} \gg D_\mathrm{B}$, then it is expected that the markers will remain at the $\mathrm{A}_n\mathrm{B}_m/\mathrm{A}$ phase boundary, while the product layer grows at the $\mathrm{B}/\mathrm{A}_n\mathrm{B}_m$ boundary. Whether it is useful in this case to speak of a "Kirkendall effect" in the sense of section 7.1.2 is debatable. It is observed experimentally, however, that in such a case the continual removal of A atoms from the $\mathrm{A}_n\mathrm{B}_m/\mathrm{A}$ phase boundary leads to the formation of pores, so that ideal contact is no longer maintained. The assumption that local thermodynamic equilibrium is maintained at the phase boundaries, which is the assumption upon which the above calculations are based, will then no longer hold true. Therefore, when performing such experiments for the purpose of testing the theory, one should apply external pressure in order that the contact between the individual phases will remain ideal.

The second possible situation which could occur for this type of reaction is the following: 2. The metal atoms A and B in the reaction product A_nB_m occupy the same lattice, just as in the case of a substitutional solution, or else the sublattices are energetically so similar that both types of atoms can move in every sublattice. Then, in contrast to the case just discussed, the fluxes j_A and j_B are once again coupled, since the condition of site-balance for an observer in the external system must be satisfied. The condition is the same as that used to derive the Darken equation in section 5.5.3. Accordingly, a general diffusion coefficient for A and B is obtained which is a combination of the component diffusion coefficients as given by the Darken equation, just as in the case of the formation of ionic crystals or in the case of diffusion in simple metallic systems. The calculations which were just performed above may now be repeated with the condition:

$$j_A = -j_B = -(N_A\bar{D}_B + N_B\bar{D}_A)\frac{c_A}{RT}\frac{d\mu_A}{dx} \tag{7-31}$$

and with eq. (7-27) now modified to read:

$$\frac{d\Delta x}{dt} = \frac{n+m}{nm} V_m j_A \tag{7-27a}$$

The result of the calculations is, as expected, that the growth rate law for the product layer and the expression for the practical parabolic rate constant $\bar{k}$ are exactly the same as in eqs. (7-28) and (7-30). The result could not be otherwise, since the expression for the same phenomenological process, described in terms of purely phenomenological quantities, cannot depend upon the particular atomic mechanism.

To conclude this section we shall discuss the case in which a multiphase product layer is formed between two reactants A and B. We shall assume that the range of homogeneity of each phase is sufficiently narrow so that the above treatment will apply to the individual phases. We may then proceed on the assumption that the parabolic reaction rate constants for each individual phase are known. In this case, however, $\Delta\mu_A$ is no longer given by eq. (7-29), where $\Delta G^0_{A_nB_m}$ is the free energy of formation of A_nB_m from the elements A and B, but rather, $\Delta\mu_A$ is given as follows. To be completely general, let us consider that the reaction product consists of a series of phases $A_{n_1}B_{m_1}$, $A_{n_2}B_{m_2}$, etc. The following equilibrium conditions then apply at the phase boundaries:

$$\nu_1 A_{n_1}B_{m_1} + A = \varrho_2 A_{n_2}B_{m_2}$$

$$\nu_2 A_{n_2}B_{m_2} + A = \varrho_3 A_{n_3}B_{m_3} \quad \text{etc.}$$

The difference $\Delta\mu_A$ between the two phase boundaries of the p'th reaction product layer is then given as:

$$\Delta\mu_A^{(p)} = \nu_{p-1}\mu^\circ_{A_{n_{p-1}}B_{m_{p-1}}} - (\varrho_p + \nu_p)\mu^\circ_{A_{n_p}B_{m_p}} + \varrho_{p+1}\mu^\circ_{A_{n_{p+1}}B_{m_{p+1}}} \tag{7-32}$$

That is, the $\Delta\mu_A^{(p)}$ are known if the thermodynamics of the reacting system A-B is known.

The rest of the treatment of this problem follows the lines of section 6.2.3. In particular, it should be noted that, according to eq. (6-27), the total thickness of the multiphase intermetallic product layer obeys a parabolic growth rate law in a one-dimensional experiment as long as the assumption of equilibrium at the phase boundaries holds true. In section 6.2.3, no special reference was made to the nature of the reacting system. Therefore, the relationships derived in section 6.2.3 apply equally well to the present case, and so section 6.2.3 should be read over once again in detail. A great number of examples of the type of system discussed here can easily be found, as one can appreciate by simply glancing at a collection of metallic phase diagrams. Special mention should be made of the systems U-Al, U-Co, and Zr-Al which are important in reactor technology. A great many qualitative studies have been made in this area, and since the advent of the microprobe, a number of quantitative studies have also been made. Unfortunately, however, a complete set of thermodynamic and kinetic data which would permit the rates of reactions of the type discussed in this section to be calculated in advance is not yet available [25, 26].

Finally, the anisotropy which is found in many intermetallic systems should be briefly discussed. In the derivation of the kinetic equations it was tacitly assumed that the reactants and products are isotropic. For many intermetallic systems, however, this is not the case. The formulae then apply only to polycrystalline material, and the "average" transport coefficients are averages with respect to the directional dependence as well as to the activity dependence of these coefficients. In this case, too, a certain unevenness of the phase boundary will arise. However, if the solid state reaction occurs between two coarse-grained substances or between metallic single crystals (in which latter case a monocrystalline product will also be formed under favourable conditions), then the complete matrix of the diffusion coefficient tensor must be used in the calculations. A pronounced columnar growth is then observed. An example is the formation of the tetragonal η phase Fe_3Al_5 by the reaction between Fe and Al. In this case, diffusion along the c-axis of the reaction product predominates because of the structural situation, so that a growth texture results [27].

7.2.3. Formation of intermetallic phases with extended ranges of homogeneity

The reactants are once again two pure metals A and B. In Fig. 7-5 is shown a possible concentration profile after annealing at diffusion temperatures. The corresponding phase diagram is also given here. Two phases, β and γ, with finite ranges of homogeneity are formed from the end phases α and δ. Because it is based upon completely general assumptions, the Boltzmann-Matano analysis can also be used in the present case of a multiphase reaction layer to determine the chemical diffusion coefficient $\tilde{D}$ which suffices to give a phenomenological description of the diffusional process in the case of binary diffusion [25]. Eq. (5-54) was extended in eq. (7-12) so as to permit the chemical diffusion coefficient to be calculated from the measured concentration profile in systems with variable molar volumes. This equation may be further modified to apply to a multiphase system [20] as follows:

$$\tilde{D}(N_2^{(k)}) = \frac{1}{2\,t\left(\dfrac{\partial N_2^{(k)}}{\partial x}\right)}\left[\frac{N_2^+ - N_2^{(k)}}{N_2^+ - N_2^-}\int\limits_{-\infty}^{x(k-1,k)} \frac{V_m^{(k)}}{V_m}(N_2 - N_2^-)\,\mathrm{d}x + \frac{N_2^{(k)} - N_2^-}{N_2^+ - N_2^-}\int\limits_{x(k,k+1)}^{+\infty} \frac{V_m^{(k)}}{V_m}(N_2^+ - N_2)\,\mathrm{d}x + \frac{(N_2^{(k)} - N_2^-)(N_2^+ - N_2^{(k)})}{N_2^+ - N_2^-}\,\Delta x^{(k)}\right] \qquad (7\text{-}33)$$

(As usual, $N_2^{(k)}$ indicates a value of N_2 in phase k.) $x\,(k-1, k)$ and $x\,(k, k+1)$ are the coordinates of the phase boundaries between phases k-1 and k and between phases k and $k+1$ respectively. $\Delta x^{(k)}$ is the thickness of phase k, and $V_{\mathrm{m}}^{(k)}$ is the molar volume of phase k. For phases with more or less extended ranges of homogeneity, eq. (7-33) can be applied with no difficulty. If the phase k possesses a very narrow range of homogeneity, however, then it is practically impossible to measure $\partial N_2^{(k)}/\partial x$ experimentally with any accuracy, and it is no longer practicable to use eq. (7-33). The procedure which should be followed in such cases has already been discussed in detail in section 7.2.2.

In practice, it is especially important to be able to determine in a simple way the average chemical diffusion coefficients $\overline{\tilde{D}^{(k)}}$ for the individual phases from the concentration profiles. These concentration profiles can be easily measured nowadays through the use of the microprobe. $\overline{\tilde{D}^{(k)}}$ is defined as:

$$\overline{\tilde{D}^{(k)}} = \frac{1}{\Delta N_2^{(k)}} \int\limits_{N_2\,(k-1,\,k)}^{N_2\,(k,\,k+1)} \tilde{D}\,(N_2)\,\mathrm{d}N_2 \tag{7-34}$$

where $\Delta N_2^{(k)} = N_2\,(k, k+1) - N_2\,(k-1, k)$, and where $N_2\,(k, \mathrm{k}+1)$ and $N_2\,(k-1, k)$ are the mole fractions of component 2 in phase k at the right-hand and left-hand phase boundaries respectively. If pure metals A and B are used as starting substances, then $N_2^- = 0$, and $N_2^+ = 1$. By substituting eq. (7-33) into eq. (7-34) and carrying out the integration, it can be shown that:

$$\overline{\tilde{D}^{(k)}} = \frac{\Delta x^{(k)}}{2\,t\,\Delta N_2^{(k)}} \Bigg[\Delta x^{(k)} N_2^{(k)} (1 - N_2^{(k)}) + (1 - N_2^{(k)}) \int\limits_{-\infty}^{x\,(k-1,\,k)} \frac{V_{\mathrm{m}}^{(k)}}{V_{\mathrm{m}}} N_2\,\mathrm{d}x + N_2^{(k)} \int\limits_{x\,(k,\,k+1)}^{+\infty} \frac{V_{\mathrm{m}}^{(k)}}{V_{\mathrm{m}}} (1 - N_2)\,\mathrm{d}x \Bigg] \tag{7-35}$$

Here, $N_2^{(k)}$ is an average value of N_2 in phase k which is assumed to have a narrow range of homogeneity. If all the individual phases exhibit relatively narrow regions of homogeneity, then we can go one step further and carry out the integrations assuming constant fixed average values for the $V_{\mathrm{m}}^{(k)}$ as well. In this way we obtain the following expression:

$$\overline{\tilde{D}^{(k)}} = \frac{\Delta x^{(k)}}{2\,t\,\Delta N_2^{(k)}} \Bigg[\Delta x^{(k)} N_2^{(k)} (1 - N_2^{(k)}) + (1 - N_2^{(k)}) \sum_{v=2}^{k-1} \frac{V_{\mathrm{m}}^{(k)}}{V_{\mathrm{m}}^{(v)}} N_2^{(v)} \Delta x^{(v)} + N_2^{(k)} \sum_{v=k+1}^{n-1} \frac{V_{\mathrm{m}}^{(k)}}{V_{\mathrm{m}}^{(v)}} (1 - N_2^{(v)}) \Delta x^{(v)} \Bigg] \tag{7-36}$$

where n is the total number of phases present. The application of eq. (7-36) is very simple if sufficiently accurate concentration profiles are available. It must always be kept in mind that the equations of this section are only applicable if ideal contact and thermodynamic equilibrium are achieved at all phase boundaries. It can be appreciated that as the number of phases present increases, the probability that these conditions will be met decreases correspondingly. Deviations from the parabolic rate law and the formation of pores during solid state reactions have been observed many times [28]. These observations can be of great interest from the technological standpoint, as for example, in connection with cold welding and hot welding.

If one wishes to examine the reaction mechanism or to determine transport coefficients in an intermetallic phase $A_n B_m^{(k)}$, then, on the basis of the preceding discussion it would seem advisable to form the diffusion couple from the two saturated neighbouring phases $(k-1)$ and $(k+1)$, and not from the pure metals A and B. The rate of growth of the k'th phase will not be the same for these two cases, since in the latter case (pure A and B as starting substances) the kinetic and thermodynamic parameters of all the phases in the diffusion couple will play a role in determining the growth rate of phase k, whereas in the former case only the parameters of phase k itself will take part. Once again, a distinction can be made between reaction constants of the first and second kind (see section 6.2.3).

The following equations define rate constants of the first and second kind:

$$\bar{k}_{\mathrm{I}} = \frac{\Delta x_{\mathrm{I}}^2}{2\,t}\,; \qquad \bar{k}_{\mathrm{II}} = \frac{\Delta x_{\mathrm{II}}^2}{2\,t}$$

Δx_{I} is the reaction layer thickness in the case where it is formed in a reaction starting with the pure reactants. Δx_{II}, on the other hand, is the reaction layer thickness in the case of a reaction where the reaction layer under consideration is formed from the saturated adjacent phases. These definitions hold for a one-phase reaction product as well as for a multiphase reaction product. Since the average diffusion coefficients in a certain reaction layer (phase k) cannot depend upon the starting material if local equilibrium prevails, one may return to eqs. (7-35) and (7-36) in order to obtain a relation between the reaction rate constants of the first and second kind. By taking into account the definitions, the relation between the two rate constants for a reaction product with phases of very narrow ranges of homogeneity can be shown to be:

$$\bar{k}_{\mathrm{II}}^{(k)} = \bar{k}_{\mathrm{I}}^{(k)} \cdot \frac{N_2^{(k+1)} - N_2^{(k-1)}}{(N_2^{(k)} - N_2^{(k-1)})\,(N_2^{(k+1)} - N_2^{(k)})} \cdot \Bigg[(1 - N_2^{(k)}) \cdot N_2^{(k)} + \frac{(1 - N_2^{(k)})}{\Delta x^{(k)}} \sum_{\nu=2}^{k-1} \frac{V_{\mathrm{m}}^{(k)}}{V_{\mathrm{m}}^{(\nu)}} \cdot N_2^{(\nu)} \cdot \Delta x^{(\nu)} + \frac{N_2^{(k)}}{\Delta x^{(k)}} \cdot \sum_{\nu=k+1}^{n-1} \frac{V_{\mathrm{m}}^{(k)}}{V_{\mathrm{m}}^{(\nu)}} \cdot (1 - N_2^{(\nu)}) \cdot \Delta x^{(\nu)}\Bigg] \tag{7-36a}$$

Eq. (7-36a) gives the relation between the rate constants of the first and second kind if the diffusion profile is known as a function of the distance coordinate, or if the thickness and composition of the product phases are known. Similar equations were first derived by Wagner [20]. The most important conclusion for a one-phase reaction product (phase (2)) is deduced when phase (1) is the one reactant, phase (3) is the other reactant, and the index s designates the saturation concentration of phase (1) in phase (3) and vice versa. Then:

$$\left(\frac{\bar{k}_{\mathrm{II}}}{\bar{k}_{\mathrm{I}}}\right) \simeq \frac{(1 - N_2) \cdot N_2 \cdot (N_{2,\mathrm{s}}^{(3)} - N_{2,\mathrm{s}}^{(1)})}{(N_2 - N_{2,\mathrm{s}}^{(1)})\,(N_{2,\mathrm{s}}^{(3)} - N_2)} \cdot \Bigg[1 + \frac{V}{\Delta x}\Bigg\{\frac{1}{V^{(1)} \cdot N_2} \cdot \int^{\text{phase (1)}} N_2^{(1)}\,\mathrm{d}x + \frac{1}{V^{(3)}(1 - N_2)} \cdot \int^{\text{phase (3)}} (1 - N_2^{(3)})\,\mathrm{d}x\Bigg\}\Bigg] \tag{7-36b}$$

where terms without superscript refer to the reaction product, and the integrals extend over the respective phases. For negligible solubility in the reactants, the integrals vanish, and $\bar{k}_{\mathrm{II}} = \bar{k}_{\mathrm{I}}$. In any case, $\bar{k}_{\mathrm{II}} \geqslant \bar{k}_{\mathrm{I}}$ for eq. (7-36b). Eq. (7-36b) is useful if, for example, the diffusion profile and Δx have been measured by electron-microprobe analysis after a solid state reaction of the first kind in order to obtain the more fundamental reaction rate constant of the second kind, from which the transport coefficients can then immediately be calculated if the thermodynamics of the A-B system is known.

An example of the formation of intermetallic phases of the aforementioned type can be found in reference [44]. It must be stressed, however, that the foregoing analysis is in no way restricted to metallic phases, and can be applied in semiconductor and ionic systems as well.

7.3. Precipitation phenomena in supersaturated solid solutions

Reactions in metallic systems in which a new phase precipitates from a supersaturated solution have very broad technological applications. For instance, such processes cause the precipitation hardening of alloys. The yield point and the strength of the alloy are increased as a result of the interaction of the dislocations with the precipitated particles. Accordingly, the quantitative treatment of a hardening problem requires that the number, size, distribution, and shape of the precipitated particles be known. In the following sections, certain kinetic problems which play an important role in this regard will be discussed [8].

7.3.1. Thermodynamic and structural aspects of precipitation

In order for precipitation phenomena to take place, a supersaturation must occur. This will, in general, arise as a result of the undercooling of a homogeneous solid solution into a two-phase region (i. e. as a result of the crossing of a solubility line). Further discussion will be based upon the situation shown in Fig. 7-8.

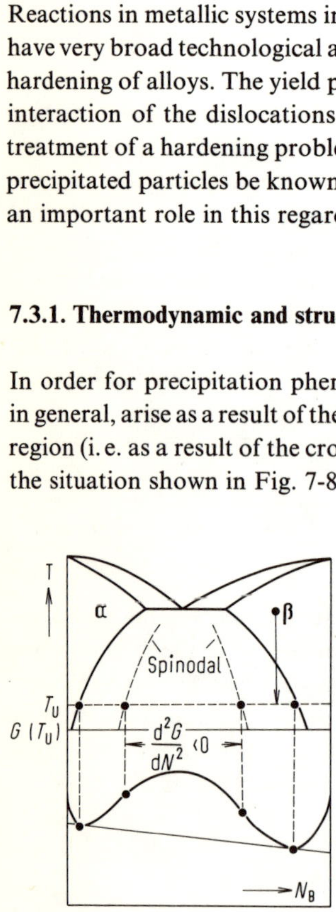

Fig. 7-8. Schematic presentation of the phase diagram and the free energy of the solid solution when a miscibility gap occurs. T = temperature, N_B = mole fraction of B, $G(T_U)$ = free energy at the temperature of precipitation T_U.

In Fig. 7-8, the Gibbs free energy of the undercooled solution is shown as a function of composition. The common tangent line as drawn on this diagram gives the position of the miscibility gap on the $T - N_i$ diagram as a result of the relationship $(\partial G^M/\partial N_i)_{P,T,j \neq i} = \mu_i$. The spinodal curve separates the region of spontaneous demixing from the region in which an activation is required before phase separation can occur. Here the nucleation energy which must always be supplied due to the formation of a new interface is disregarded (see also section 7.4). Of course, an equilibrium diagram does not supply complete information, inasmuch as considerable distortions of the lattice can occur during the formation of the new phase in a solid matrix. These distortions arise because of the difference in the molar volumes of the solid solution and the precipitating phase. This is not evidenced in the normal free energy diagram. In theoretical treatments, the energy fraction arising from these distortions is taken to be proportional to the volume of the nucleus. If these effects are neglected (i. e. if only those cases are considered for which the change in molar volume is sufficiently small), then it is very easy to calculate the free energy $\Delta g^{N}(r^*)$ which is required for the formation of a nucleus of the precipitating phase of critical radius r^* so that the nucleus is stable and is capable of growing [9]. This energy is given by:

$$\Delta g^{N}(r^*) = \frac{1}{3}(4\pi r^{*2}\gamma_0) \tag{7-37}$$

That is, 1/3 of the total interfacial free energy $4\pi r^{*2}\gamma_0$ (γ_0=specific interfacial free energy) must be supplied in order to form a stable nucleus. The remaining 2/3 is recovered when the atoms leave the bulk of the supersaturated phase and precipitate as the new phase, since the precipitating phase is the thermodynamically stable phase. That is, the bulk free energy of the precipitated phase is more negative than that of the supersaturated solution. All that has been said up to now applies only for the case of homogeneous nucleation. If crystal defects such as dislocations and grain boundaries are present, then the treatment must be modified, since when a new phase is formed on these crystal defects, the energy of the defect will then be added to the (negative) volume energy of the nucleus. For such cases of so-called heterogeneous nucleation, this can result in a situation in which every nucleus can grow without first having to attain a critical size r^*. Because of the presence of dislocations, low-angle grain boundaries, normal grain boundaries, surfaces, and inclusions in crystalline solids, the usual mode of nucleation is found to be the heterogeneous nucleation, which is difficult to assess in a quantitative way.

Furthermore, the interfacial energy at the commencement of growth of the new phase will be strongly dependent upon whether the nucleation is fully coherent, partially coherent (nuclei formation by shear), or completely incoherent. The meaning of these terms is schematically illustrated in Fig. 7-9.

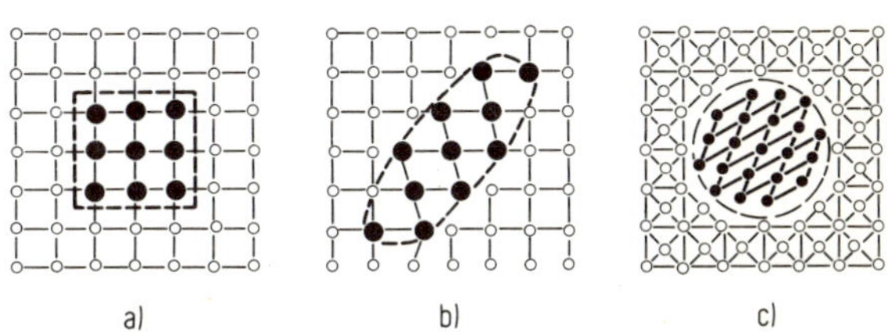

Fig. 7-9. The various transitions between nucleus and matrix: (a) coherent, (b) partially coherent, (c) incoherent (see also reference [32]).

Most precipitation phenomena appear to proceed either coherently or partially coherently, at least in their very early stages. This can be inferred from the continual recurrence of certain relationships in the orientations between the supersaturated matrix and the precipitated phase. An example is an orientational relationship named after Kurdjumov and Sachs which is frequently found for the precipitation of ferrite in austenite. This relationship is: $(111)_{\text{austenite}} \| (110)_{\text{ferrite}}$. That is, the close-packed planes of the face-centered-cubic and body-centered-cubic lattices are parallel to one another [10].

If the nucleation is a thermally activated process, then the rate of nucleation will depend exponentially upon the free energy of formation of a nucleus of the critical size for growth to occur. The Gibbs-Thomson relation:

$$\Delta \mu = \frac{2\, V_{\mathrm{m}} \gamma_0}{r}$$

gives the change in chemical potential of matter in small particles of radius r relative to a planar interface. By using this relationship along with eq. (7-37), it can be shown that the rate of nucleation varies exponentially as the third power of the specific interfacial free energy γ_0. This strong dependence of the nucleation rate upon γ_0 explains why the supersaturated phase seeks a mode of nucleation involving the lowest possible surface energy in order to decrease the supersaturation in the shortest possible time. This may be achieved by incorporation of dislocations and grain boundaries (heterogeneous nucleation) or by the adoption of favourable orientations relative to the matrix. In seeking numerical values for the interfacial free energy, one should keep in mind that the everpresent impurities in the metal will, to some extent, tend to collect preferentially at the interfaces, with an excess surface concentration Γ_i given by the Gibbs adsorption isotherm:

$$\Gamma_i = -\,(\mathrm{d}\gamma_0/\mathrm{d}\mu_i)_{T,j \neq i}$$

This can influence the value of γ_0 considerably.

The first stages of the formation of a new phase, as discussed above, determine to a large extent its spatial configuration and distribution. In the following sections, the reaction kinetics applicable to the growth of a newly formed phase in a supersaturated matrix will be treated.

7.3.2. Growth kinetics of precipitation

A very early and very clear quantitative treatment of the growth kinetics of precipitated particles in a supersaturated matrix was given by Zener [39] for a one-dimensional problem. The treatment which we shall present here is basically the same as Zener's, although the details are somewhat different. Let us assume that, at time $t = 0$, there exists, per unit volume of supersaturated matrix, a known number z of nuclei of the precipitating phase which are capable of further growth. These nuclei are assumed to be distributed completely randomly throughout the volume of the matrix. Each nucleus is assigned a spherical region of influence. The average radius r_{e} of these regions is given by the simple relation:

$$\frac{4}{3}\pi r_{\mathrm{e}}^3 z = 1; \quad r_{\mathrm{e}} = \left(\frac{3}{4\,\pi z}\right)^{1/3} \tag{7-38}$$

The problem now is to solve Fick's second law, expressed in spherical coordinates, under the assumption that: 1. The concentration of the precipitating component at time $t = 0$ is everywhere equal to c^0; 2. For $t > 0$, thermodynamic equilibrium is always maintained at the phase boundary between matrix and precipitate, where the precipitate is approximated as a sphere of radius r_p (i.e. at the phase boundary, $c = c\,(\text{eq})$); and 3. No mass flux occurs across the boundaries of the individual regions of influence (i.e. $(\partial c/\partial r)_{r_e} = 0$). If Fick's second law is written in spherical coordinates for the case of a constant chemical diffusion coefficient as follows:

$$\frac{\partial c}{\partial t} = \tilde{D}\left[\frac{\partial^2 c}{\partial r^2} + \frac{2}{r}\frac{\partial c}{\partial r}\right] \tag{7-39}$$

then an approximate solution which holds for long times can be immediately written. This approximation applies when the precipitating component in the matrix crystal has almost achieved its equilibrium concentration $c(\text{eq})$ so that the radius r_p of the spherical particles of precipitate can be reasonably assumed to be constant. The details of the calculation can be found in the literature [29]. The result is:

$$\frac{\bar{c} - c(\text{eq})}{c^\circ - c(\text{eq})} = \exp - \frac{3\, r_p \tilde{D}}{r_e^3} \cdot t \tag{7-40}$$

Thus, a relaxation time $\tau = r_e^3/3\, r_p \tilde{D}$ for the precipitation process can be given. $\bar{c}$ is the average concentration of the precipitating component in the region of influence $4/3 \cdot \pi r_e^3$, and r_p is small compared to r_e. Furthermore, under the conditions of applicability of eq. (7-40) the following mass balance can be written:

$$\frac{4}{3}\pi r_e^3 \,(c^\circ - c(\text{eq})) \simeq \frac{4}{3}\pi r_p^3 c(\text{p}) \tag{7-41}$$

where $c(\text{p})$ is the concentration of the precipitating component in a particle of precipitate. The relaxation time is thus given by:

$$\tau = \frac{r_e^2}{3\,\tilde{D}}\left(\frac{c(\text{p})}{c^\circ - c(\text{eq})}\right)^{1/3} \tag{7-42}$$

In this equation are found only quantities which are, in principle, either known or easily measurable. Despite the fact that r_e is calculated by the expression in eq. (7-38), the approximation is actually quite good, since the resistance to diffusion lies essentially in the immediate surroundings of the precipitate on account of the spherical symmetry of the problem. Therefore, an exact knowledge of the distribution of the precipitate particles is irrelevant for that part of the precipitation process being considered here. For the initial stages of the precipitation, another solution of eq. (7-39) must be sought, since here the radius r_p of the precipitate particles can no longer be taken to be independent of time. It can be seen immediately that the following mass balance now replaces that given in eq. (7-41):

$$\frac{4}{3}\pi r_e^3 \,(c^\circ - \bar{c}) \simeq \frac{4}{3}\pi r_p^3(t)\, c(\text{p}) \tag{7-43}$$

$r_p(t)$ is the radius of the precipitate particles at time t. Furthermore, the instantaneous conservation equation for the precipitating component may be written in the form:

$$\frac{4}{3}\pi r_e^3 \frac{d\bar{c}}{dt} \simeq -\tilde{D}\left(\frac{\partial c}{\partial r}\right)_{r_p} \cdot 4\pi r_p^2 \tag{7-44}$$

This equation says that the total flux into the spherical precipitate particle with radius r_p at time t causes a decrease in the average concentration $\bar{c}$ in the region of influence. As long as $\bar{c}$ has not changed much from c°, a quasi-steady state is set up in the neighbourhood of the matrix/precipitate phase boundary. This quasi-steady state condition simplifies Fick's second law (7-39) considerably, so that a direct integration is possible. The result is of the form $c \propto 1/r +$ + constant. By combining this result with the conditions in eqs. (7-43) and (7-44), it can be easily shown, after some elementary rearrangement, that the average concentration $\bar{c}$ in the region of influence is given by

$$\bar{c} = c^\circ - \left[\frac{[2\,\tilde{D}\left(c^\circ - c(\mathrm{eq})\right)]^3}{c(\mathrm{p})\,r_e^6}\right]^{1/2} t^{3/2} \tag{7-45}$$

Once again, all the factors in this equation are, in principle, known or measurable, so that the initial stages of precipitation can also be treated quantitatively. For a more detailed discussion of the range of applicability and the error limits of these results, reference should be made to the special literature [30]. In every case, one must keep in mind the limiting assumptions which were used to obtain eqs. (7-42) and (7-45). These are: the matrix and precipitate have equal molar volumes; all effects connected with the surface free energy are neglected; the geometrical shape of the growing phase boundaries remains stable. The fact that dendritic precipitates can often be observed indicates that this last assumption must be examined more closely. This will be the subject of the next section.

First of all, however, in light of the preceding general discussion, let us consider, as an example, the precipitation of cementite Fe_3C out of an iron matrix which is supersaturated with carbon [40]. Difficulties arise because of the difference between the molar volume of the precipitate and the volume per mole of the precipitating components in the matrix. This causes a stress field to be formed around the precipitate. As long as no internal voids or creep processes occur, in the steady state case there will be a coupled diffusional flux of carbon towards the precipitate and of iron towards the supersaturated matrix, because the volume of the Fe_3C precipitate is greater than the volume of 3 Fe. This problem was first analyzed by Hillert [41] for the system Fe-C. This analysis, however, has a quite wide and general applicability.

In current thermodynamic terminology this rather complicated process can be related to the pressure dependence of the partial free energies. It is the gradients of these partial free energies which are the driving forces for the component fluxes. With the Gibbs-Thomson equation which has already been given, and with the generalized Gibbs-Duhem equation in the form $\sum N_i \, d\mu_i = V_m \, dP$, we have at our disposal a complete system of equations for calculating the kinetics of precipitation. The calculations are elementary, but rather lengthy. Those who are interested are referred to the original literature [41, 42, 43].

The growth of an Fe_3C particle in a supersaturated austenite matrix can be described on an atomistic basis as follows. The rate-determining step is the relatively slow diffusion of iron via vacancies. The flux of iron occurs away from the surface of the cementite particles towards the matrix. The main contribution to the driving force is the pressure difference between the surface of the cementite and the interior of the matrix. The reasons for this are, firstly, that the vacancy concentration in the iron sublattice is lowest in the neighbourhood of the precipitate on account of the higher pressure which is prevalent there. This will promote a motion of the iron atoms in the direction of the pressure gradient. Secondly, the interstitial diffusion of the carbon towards the precipitate will be retarded as a result of the increased pressure. Together, these effects give rise to the coupled flux of iron and carbon in the steady state as discussed above.

7.3.3. Stability of moving phase boundaries

If local thermodynamic equilibrium is maintained at the phase boundary between the supersaturated matrix and the precipitating spheroids, as has been continually assumed in the preceding sections, then, in the absence of further stabilizing factors which have not yet been discussed, a small disturbance in the spherical symmetry of the growing phase boundary must lead to an instability of the spherical shape and to a dendritic type of growth. This can be easily seen if iso-concentration lines are drawn around such a disturbance as shown in Fig. 7-10. The density of these lines is a measure of the concentration gradient, which can be seen to be greater beside a point where there is a bowing-out of the precipitating particle into the matrix than it is beside an indentation in the particle. Since the mass flux is proportional to this concentration gradient, the flux towards the projection will be greater than that towards the indentation. The projection will thus grow, until it finally degenerates into a needle. What, then, are the stabilizing factors that can account for the fact that, in most cases, growing phase boundaries are observed to remain stable during precipitation?

Fig. 7-10. Model of a moving phase boundary during the precipitation of the α-phase out of a supersaturated β-matrix. A sinusoidal disturbance of the planar surface is shown with iso-concentration lines: (a) without consideration of interface tension, and (b) when interface tension is considered. The possibility of mass transport along the phase boundary is also indicated.

The geometrical shape of moving phase boundaries, as well as that of stationary boundaries, is influenced substantially by the surface or interface free energy. For sufficiently mobile particles, this can be regarded as a surface tension [11]. According to the Gibbs-Thomson equation, there will be a change in chemical potential at a curved phase boundary relative to a planar phase boundary. The difference will become greater as the corresponding radius of curvature becomes smaller. For example, let us assume that, in an originally planar surface, a sinusoidal disturbance $y = \varepsilon(t) \sin \omega x$ occurs during growth, which is cylindrical in the z-direction. The chemical potential difference between the projection and the indentation for the case of precipitation of an α-phase consisting of virtually pure substance A is then:

$$\Delta \mu_A = 2\, V_A \gamma_0\, \varepsilon(t) \omega^2 \tag{7-46}$$

Eq. (7-46) can be derived by calculating the curvature of the surface, and then substituting this into the Gibbs-Thomson equation. The potential difference $\Delta \mu_A$ is thus proportional to the instantaneous amplitude $\varepsilon (t)$ of the sinusoidal disturbance, and inversely proportional to the square of the wave length $\lambda = 2 \pi/\omega$. This, then, leads to an evening out of the iso-activity lines as is shown in Fig. 7-10b. It may be easily seen from eq. (7-46) that there is a certain limiting wave length λ_{max}. For wave lengths less than this, disturbances of amplitude ε will decay, but for greater wave lengths, these disturbances will increase in size, and the growing planar interface will become unstable. This limiting wave length λ_{max} brings the activity of A at the phase boundary to the level of supersaturation in the matrix.

The numerical value of λ_{max} will depend upon the concentration gradient $\mathrm{d}c_A/\mathrm{d}y$ in the vicinity of the precipitate. This concentration gradient will naturally decrease as the precipitation proceeds, so that λ_{max} will continually increase with time. Finally, in the limiting case, a completely stable phase boundary will be achieved.

Only the simplest fundamental aspects of the stability of moving interfaces have been discussed here. For instance, we have not considered the transport of substance A from sites of maximum projection to sites of maximum indentation by means of diffusion along the interface or through regions near the interface. When the interfacial diffusion coefficients are large, this process can also produce a large stabilizing effect. We have also not considered the influence of a slow phase boundary reaction (i.e. a slow step involving the transfer of A out of the supersaturated matrix into the particle of precipitate at the phase boundary). This barrier to transfer could be rate-controlling during the initial stages of the precipitation process when the diffusional fluxes are large as a result of steep concentration gradients. This would then mean that the overall precipitation process is no longer diffusion-controlled as is implicit in the assumption of local thermodynamic equilibrium at the phase boundary. Whether certain geometrical shapes are stable or not would then depend upon the rate of transfer of particles across certain crystallographic planes, and upon the surface free energy. Also, the kinetic equations given in the previous section were derived under the assumption that the precipitation process is diffusion-controlled. These equations would have to be modified if the overall process were reaction-controlled. Finally, it should be mentioned that distortions can arise in the crystal because of the difference between the molar volumes of the growing precipitating particles and the solid solution. If these distortions are not removed quickly enough by the action of plastic deformation or by the formation or destruction of lattice points by means of vacancy diffusion to or from sites of repeatable growth in the crystal, then the directional dependence of the distortions can induce growing particles of precipitate to take on certain specific shapes.

7.3.4. Ageing and Ostwald ripening

Let us assume that precipitation of the A-rich α-phase has occurred to the extent that the matrix is no longer supersaturated, and that we now have a two-phase alloy which is nearly at equilibrium. In section 7.3.2, it was assumed that the precipitate particles were all of the same size. In reality, however, there will be a range of particle sizes, since not all nuclei were formed at the same time, and also since the supply of A to a precipitate particle will be influenced by the site of the particle in the matrix. According to the Gibbs-Thomson relation, smaller particles of precipitate have a higher chemical potential for component A than do the larger particles. For spherical particles of radius r' and r'', the difference in chemical potential is:

$$\Delta \mu_A = 2\, V_m \gamma_0 \frac{r' - r''}{r' r''} \tag{7-47}$$

Thus, the larger particles will grow at the expense of the smaller particles, thereby decreasing the total surface or interface free energy. This solid state reaction is known as Ostwald ripening. In principle, the kinetics of this process can be calculated if one knows the reaction rate constants, the transport coefficients or chemical diffusion coefficients, the distribution curve for particle size, and the geometrical arrangement of the particles in the matrix. It can easily be seen that the occurrence of Ostwald ripening in precipitation-hardening alloys will have an effect upon the mechanical properties of the material, since the yield point and other properties which depend upon the interaction between dislocations and precipitate particles will be directly influenced. The hardness decreases, and we say that the alloy is overaged.

In the quantitative treatment of Ostwald ripening, two limiting cases may be distinguished. In the first case, the rate is controlled by diffusion between the particles. In the second case, the phase boundary reaction is rate-determining. If it is assumed that the A particles obey the laws of ideal dilute solutions in the matrix, then eq. (7-47) may be rewritten in the form of a concentration difference:

$$\frac{\Delta c_A}{c_A\,(\mathrm{eq})} = \frac{2\, V_m \gamma_0}{RT} \cdot \frac{r' - r''}{r' r''} \tag{7-48}$$

The relationship $\Delta \mu_A = \Delta RT \ln c_A \approx RT \Delta c_A / c_A\,(\mathrm{eq})$ has been used in this derivation. Eq. (7-48) says that there is a relative concentration gradient between two precipitating particles of radius r' and r''. In the case of a diffusion-controlled equilibration process, this concentration gradient leads to an increase in the average radius $\bar{r}$ of the particles of precipitate. Eq. (7-49) gives $\bar{r}$ as a function of time [33]:

$$\bar{r} = \left[\frac{8\, \gamma_0 V_m^2 c_A\,(\mathrm{eq})\, \tilde{D}}{9\, RT}\right]^{1/3} \cdot t^{1/3} \tag{7-49}$$

This rate equation presupposes that a quasi-steady state distribution function $f(r)$ exists for the particle radii r. As usual, the average particle radius $\bar{r}$ is defined as:

$$\bar{r} = \frac{\int_0^\infty f(r)\, r\, \mathrm{d}r}{\int_0^\infty f(r)\, \mathrm{d}r} \tag{7-50}$$

It seems worthwhile to comment upon the quasi-steady state distribution function $f(r, t)$ in this Ostwald ripening process. The mathematical verification of the fact that this distribution function $f(r, t)$ can be written as a product $f_1(t) \cdot f_2(r/\bar{r})$ (and $f_2(r/\bar{r})$ is a universal function even in cases where the initial distribution $f\,(r, 0)$ is of Gaussian shape and of moderate width) is rather cumbersome and must be studied from the original work [33]. However, one may conceive the shape of the quasi-steady state distribution (which has a maximum between $0 \leq r/\bar{r} < 3/2$ at $r/\bar{r} = 1.135$ and is essentially zero at $r/\bar{r} > 3/2$) by realizing that it is the interplay between the activity difference of the average activity of A in the solution matrix and of the higher activity of the small particles and the lower activity of the bigger particles that causes the distribution function to generate its quasi-steady state shape.

Of interest in eq. (7-49) is the fact that $\bar{r}$ is proportional to the cube root of the transport coefficient product (c_A (eq) $\cdot$ $\tilde{D}$), and so $\bar{r}$ is also proportional to the cube root of the solubility of A in the matrix.

If the reaction is not diffusion-controlled, but rather is reaction-controlled (i.e. if the phase boundary of the precipitate is the rate-controlling reaction barrier), then, under the same assumptions as were used to derive eq. (7-49), it can be shown that:

$$\bar{r} = \frac{8}{9}\left[\frac{\gamma_0 V_m^2 c_A(\text{eq})\, k}{RT}\right]^{1/2} \cdot t^{1/2} \qquad (7\text{-}51)$$

k is the constant of the phase boundary reaction, and is defined by the flux equation at $r = r_p$:

$$j_A = k\,(c_A(\text{eq}) - c_A) \qquad (7\text{-}52)$$

That is, the rate of transport of material into the precipitate particle is proportional to the deviation from the equilibrium concentration of A in the matrix. A similar quasi-steady state distribution function as discussed above is again obtained, its maximum value is at $r/\bar{r} = 9/8$.

It is of practical importance to know the number z of precipitate particles as a function of time. z continually decreases with time as a result of the Ostwald ripening process, since the large particles grow at the expense of the small particles. After an infinitely long time, the precipitated phase should, in theory, consist of one single compact particle. In the case of diffusion-controlled mass transport it can be shown that:

$$z = \frac{z(0)}{(1 + const \cdot t)} \qquad (7\text{-}53)$$

(where $z(0) = z(t = 0)$). In the case of reaction-control, after sufficiently long times:

$$z = \frac{z(0)}{const \cdot t^{3/2}} \qquad (7\text{-}54)$$

By means of eqs. (7-49), (7-50), (7-53) and (7-54), the kinetics of ageing can be treated quantitatively. It should also be noted that it is possible, from the time-dependence of the average particle radius, to decide whether the solid state reaction is diffusion-controlled or reaction-controlled [40].

To conclude this section, let us once again stress the technological importance of the processes just described. The properties of most technologically important metals are influenced to a large extent by phase transformations and precipitation. The morphology of the reaction product is a result of the kinetic processes of nucleation and growth, and of phase boundary effects. A consistent mathematical theory of nucleation and growth has recently been presented [31]. The problems which have been raised here are treated in depth in the special literature [10, 12].

7.4. Spinodal decomposition

In the preceding sections, the kinetics of solid state reactions have been discussed for the case in which a decrease of the supersaturation of an initially homogeneous alloy leads to the formation of a new phase with a distinct phase boundary. Fig. 7-11a gives a schematic example. However, it is found experimentally that there exists a second kind of decrease of supersaturation. In this case the decomposition proceeds coherently, and for a long time there is no sharp phase boundary between the alloy matrix and the coherent crystal region which is enriched with the supersaturated component. As is shown in Fig. 7-11b, the concentration of the crystal components varies continuously and slowly from the matrix region into the enriched region. The explanation of this type of behaviour can be read from Fig. 7-8 or 7-12. Here, the free energy of the supersaturated alloy has been plotted as a function of composition. Inside the miscibility gap, two regions can be distinguished. They are characterized by a negative or a positive curvature of the G_m (N_B)-curve, and they are separated by the spinodal line. Inside this line, the solid solution is unstable, since any fluctuation in composition leads to a decrease of the free energy of the alloy.

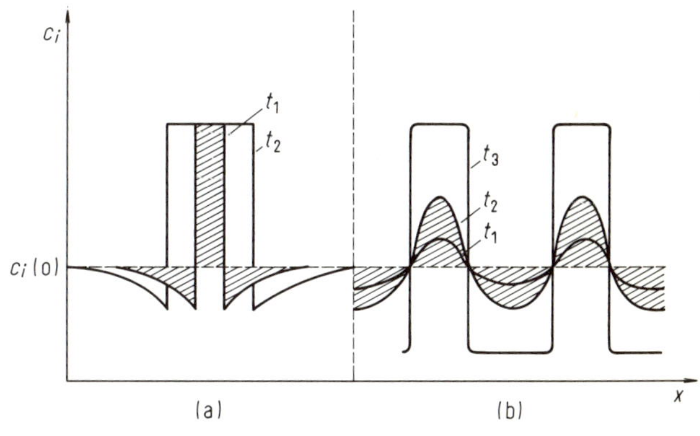

Fig. 7-11. Concentration versus distance coordinate x: (a) for the case of normal homogeneous nucleation and subsequent growth of the nucleus, and (b) for the case of spinodal decomposition. t = time. $t_3 \gg t_2 > t_1$.

On the other hand, the region between the spinodal line and the boundary of the one-phase field must be called metastable. A fluctuation of concentration in this region does not continue to grow spontaneously, as can be directly seen from Fig. 7-12. Only if a concentration fluctuation has resulted in a relatively large concentration difference in a small crystal region, can the formation of a nucleus of the new phase, as described in the section 7.3.1 on homogeneous nucleation, take place, overcoming the surface or interface free energy and lowering the overall free energy.

One might ask why the part of the free energy which is associated with the transition zone ("phase boundary"), which is present in every case, has no significant influence in the concentration range inside the spinodal lines. The answer is that, in the course of spinodal decomposition, long range concentration fluctuations occur at first with small concentration gradients. Thus, that part of the free energy which is associated with the transition zone is always smaller than the gain in volume free energy.

From this discussion of the concept of spinodal decomposition it follows that, for a description of the kinetics of the spinodal decomposition process, it is necessary to formulate

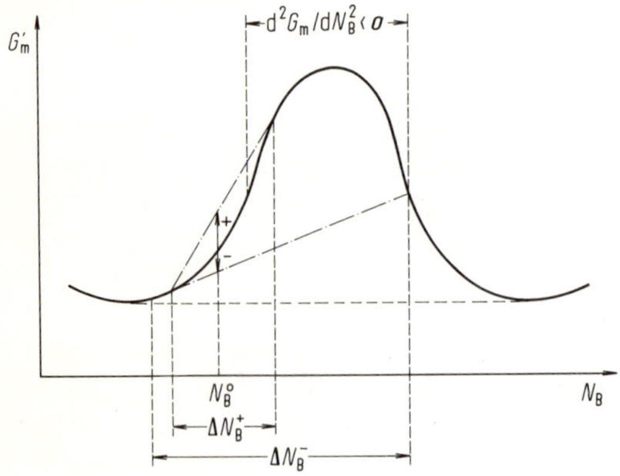

Fig. 7-12. Schematic presentation of the free energy of the solid solution versus concentration when a miscibility gap occurs.

N_B^0 = Starting concentration of homogeneous solid solution in the metastable range.

ΔN_B^+ = Concentration fluctuation starting from N_B^0, not leading to nucleation.

ΔN_B^- = Concentration fluctuation starting from N_B^0, leading to nucleation of precipitate.

the flux equations of the components, taking into account the effect of the inhomogeneity of the transition zone. In cases in which there is a significant dependence of the lattice parameter on composition, an additional strain energy must also be taken into account.

In order to allow a simple treatment, the decomposition problem will be discussed for a binary, one-dimensional system. A one-dimensional decomposition process in reality always occurs if the lattice parameter is strongly composition dependent and if the crystal exhibits elastical anisotropy. In this case, the strain energy during decomposition is strongly dependent upon the crystallographic direction in the crystal. The extension of the one-dimensional treatment to a three-dimensional case, however, offers no difficulties in principle.

The formal treatment of spinodal decomposition starts with the flux equations (5-34) which are of the form:

$$j_1 = -\frac{c_1 \bar{D}}{RT} \cdot \frac{d\mu_1}{dx}; \quad \bar{D} = N_1 D_2 + N_2 D_1$$

For the sake of simplicity, it may be further assumed that $\bar{D}$ is, to a good approximation, independent of concentration. As long as the strain energy plays no significant role in the free energy term, the chemical potential μ_i in the concentration gradient of the decomposition region is written in the form:

$$\mu_i = \bar{\mu}_i - 2K \cdot \frac{d^2 c_i}{dx^2} \tag{7-55}$$

$\bar{\mu}_i$ is the chemical potential of component i in the case of a homogeneous solid solution; the second term in eq. (7-55) stems from the fact that a particle of type i in the concentration gradient has, on the average, a different number of nearest neighbours at both its sides in the $\pm x$ direction. Irrespective of the symmetry of the pair bonding of pairs $1-1$ or $1-2$, an asymmetrical overall interaction energy is thus present which has been assumed to be proportional to the curvature of the concentration profile in the transition zone. The factor of proportionality K itself depends upon the pair interaction energies, a measure of which is the critical temperature. Introducing eq. (7-55) into the flux equations, one obtains:

$$j_i = -\frac{c_i \bar{D}}{RT}\frac{\partial}{\partial x}\left[\bar{\mu}_i - 2K\cdot\frac{\partial^2 c_i}{\partial x^2}\right] = M\left[\left(\frac{\partial^2 G_m}{\partial N_i^2}\right)\cdot\frac{dc_i}{dx} - \frac{2K}{V_m}\cdot\frac{\partial^3 c_i}{\partial x^3}\right] \quad (7\text{-}56)$$

where $M = N_i \bar{D}/RT$. Fick's second law is obtained by taking the divergence of the flux, with the result:

$$\frac{\partial c_i}{\partial t} = A\frac{\partial^2 c_i}{\partial x^2} - B\frac{\partial^4 c_i}{\partial x^4} + \text{nonlinear terms} \quad (7\text{-}57)$$

where $A = M(\partial^2 G_m/\partial N_i^2)$; $B = M \cdot 2K/V_m$. Nonlinear terms may, to a reasonable approximation, be neglected for the beginning period of the spinodal decomposition. One then finds (and verifies by substitution into eq. (7-57)) that the solution of the differential equation, in a first approximation, reads:

$$c_i(t) - c_i(\text{o}) = \exp - \left[A\cdot\left(\frac{2\pi}{\lambda}\right)^2 + B\cdot\left(\frac{2\pi}{\lambda}\right)^4\right] t\cdot\cos\frac{2\pi}{\lambda}x \quad (7\text{-}58)$$

A discussion of this solution of the kinetic problem of spinodal decomposition leads to the following conclusions: Every (approximately sinusoidal) fluctuation in composition of the supersaturated crystal will continue its spontaneous growth if the terms in the bracket of eq. (7-58) yield a negative value. If this value is positive, the fluctuations cannot grow but disappear. Setting the terms in the bracket equal to zero, one can calculate the minimum wavelength of the fluctuation which is capable of growth as:

$$\lambda_{min} = \frac{4\pi^2 B}{|A|} \quad (7\text{-}59)$$

All values $\lambda > \lambda_{min}$ lead to a spontaneous decomposition. In the case that $\lambda < \lambda_{min}$, the energy contained in the transition zone ("phase boundary") is too great so that the fluctuations cannot continue to grow. Furthermore, the value of the bracketed terms in eq. (7-58) has a minimum for $\lambda^* = \sqrt{2}\cdot\lambda_{min}$. Therefore, one would expect that, after some time, essentially only decomposition zones with the length λ^* are found in the crystal, thus giving the decomposed solid solution some sort of periodicity. This is actually found.

In the foregoing treatment, the influence of the elastic strain energy has deliberately been neglected. With the help of the theory of elastically deformed bodies, it is found that, in the case where there is a change in lattice constant with composition, the free energy has an additional term which is positive and reads:

$$G_\varepsilon = \bar{\eta}^2\cdot\frac{\bar{E}}{(1-\bar{\nu})}(c_i - c_i(\text{o}))^2 \quad (7\text{-}60)$$

where $\bar{E}$ is Young's modulus, $\bar{\nu}$ is Poisson's ratio and $\bar{\eta} = 1/a\cdot da/dc_i$ is the relative strain normalized for the unit of concentration change. Therefore, it follows from eq. (7-60) that, if decomposition results in straining the crystal, an additional term appears in the chemical potential μ_i in eq. (7-55) which is of the form $\bar{\eta}^2\cdot(2\bar{E}/(1-\bar{\nu}))(c_i - c_i(\text{o}))$. As a consequence, the constant A in Fick's second law not only contains the term $M(\partial^2 G_m/\partial N_i^2)$, but, in the situation outlined here, must be written as:

$$A = M\left[\left(\frac{\partial^2 G_m}{\partial N_i^2}\right) + \bar{\eta}^2 \cdot \frac{2\,\bar{E}}{1-\bar{\nu}}\right] \tag{7-61}$$

All further considerations and conclusions then follow lines analogous to those stated above. It is especially worth noting that when the positive strain energy is taken into account, $|A|$ is smaller than without strain energy. This results in a value λ_{min} which, according to eq. (7-59), is larger than without strain energy and, in the limiting case, can approach ∞. In this case, no decomposition can take place, although this cannot be read from the curve G_m versus composition N_B, which still shows a spinodal region.

Inserting numbers into eq. (7-58) and using an average λ_{min} value of about 10^{-6} cm, one obtains relaxation times $\tau \simeq [A \cdot (2\,\pi/\lambda_{min})^2 + B \cdot (2\,\pi/\lambda_{min})^4]^{-1}$ of the order of from one millisecond to one second, if $\bar{D}$ is assumed to be of the order of 10^{-10} cm^2/sec. (This is a typical value of the diffusion coefficient in the solid state.) From this it can be concluded that the advancement and course of spinodal decomposition is observable only under very favourable conditions. If a sample is cooled more or less slowly to room temperature, the decomposition has usually already reached its final stage. Again it should be stressed that the kinetic theory of spinodal decomposition, which has been outlined here essentially in the formulation as given by Cahn [45], does not apply to the later stages of the decomposition because of the restrictive assumptions used to simplify the calculations.

Spinodal decomposition is known in viscous liquids, in metals, in ionic crystals, and in glass. The deductions from the theory have been confirmed mainly through hardening of light metal alloys, especially in the systems Al-Zn and Al-Ag. However, in oxide solid solutions, as for example SnO_2-TiO_2, or in glass of the type Na_2O-SiO_2, decomposition processes of the spinodal type have also been found. They may prove to be of technological importance.

7.5. Literature

General Literature:

[1] W. Seith, Diffusion in Metallen, 2nd ed., Springer-Verlag, Berlin 1955.
[2] W. Jost, Diffusion in Solids, Liquids and Gases, 3rd ed., Academic Press, New York 1960.
[3] Y. Adda and J. Philibert, La Diffusion dans les Solides, Presses Universitaires de France, Paris 1966.
[4] P. G. Shewmon, Diffusion in Solids, McGraw-Hill Book Comp., Inc., New York 1963.
[5] J. Crank, Mathematics of Diffusion, Oxford University Press, Glasgow 1956.
[6] C. Wagner, Thermodynamics of Alloys, Addison-Wesley Publ. Comp., London 1952, p. 35.
[7] M. Hansen, Constitution of Binary Alloys, 2nd ed., McGraw-Hill Book Comp., Inc., New York 1958.
[8] A. Kelly and R. Nicholson, Precipitation Hardening in Progr. Mat. Sci. *5*, 143 (1954).
[9] J. W. Gibbs, Collected Works, Longmans, Green, New York 1931, p. 322.
[10] P. G. Shewmon, Transformations in Metals, McGraw-Hill Book Comp., Inc., New York 1969, p. 213.
[11] W. W. Mullins, Solid Surface Morphologies governed by Capillarity in "Metal Surfaces", Amer. Soc. Metals Monograph 1963.
[12] J. W. Christian, The Theory of Transformations in Metals and Alloys, Pergamon Press, New York 1965.

Special Literature:

[13] A. Seeger et al., J. Phys. Chem. Solids *23*, 639 (1962).
[14] P. Penning, Philips Res. Repts. *14*, 337 (1959).
[15] R. E. Hoffman, Acta Met. *6*, 95 (1958).
[16] H. Reiss, Phys. Rev. *113*, 1445 (1959).
[17] A. Smigelskas and E. Kirkendall, Trans. AIME *171*, 130 (1947).
[18] J. Schlipf, Z. Metallkunde *59*, 708 (1968); Acta Met. *14*, 877 (1966).
[19] F. Sauer and V. Freise, Z. Elektrochemie *66*, 353 (1962).
[20] C. Wagner, Acta Met. *17*, 99 (1969).
[21] A. G. Guy, Z. Metallkunde *58*, 164 (1967).
[22] L. S. Darken, Trans. AIME *180*, 430 (1949).
[23] J. S. Kirkaldy and L. C. Brown, Canad. Met. Quaterly *2*, 89 (1963).
[24] C. Wagner, J. Electrochem. Soc. *103*, 571 (1956).
[25] Th. Heumann, Z. Metallkunde *58*, 168 (1967).
[26] E. Starke and H. Wever, Z. Metallkunde *55*, 107 (1964).
[27] Th. Heumann, Z. Metallkunde *59*, 455 (1968).
[28] A. Bolk, Acta Met. *9*, 632 (1961).
[29] W. Rogalla and H. Schmalzried, Ber. Bunsenges. phys. Chem. *72*, 615 (1968).
[30] F. S. Ham, J. Phys. Chem. Solids *6*, 335 (1958).
[31] L. Kampmann and M. Kahlweit, Ber. Bunsenges. phys. Chem. *71*, 78 (1967).
[32] E. Hornbogen, Z. Metallkunde *56*, 133 (1965).
[33] C. Wagner, Z. Elektrochem., Ber. Bunsenges. phys. Chem. *65*, 581 (1961).
[34] J. S. Kirkaldy and G. R. Purdy, Canad. J. Physics *40*, 208 (1962).
[35] A. Vignes and J. P. Sabatier, Trans. AIME *245*, 1795 (1969).
[36] C. W. Taylor, M. A. Dayananda and R. E. Grace, Met. Trans. *1*, 127 (1970).
[37] R. P. Smith, Acta Met. *1*, 578 (1953).
[38] G. Tammann and H. J. Rocha, Z. anorg. allg. Chem. *199*, 289 (1931).
[39] C. Zener, J. Appl. Phys. *20*, 950 (1949).
[40] K. M. Vedula and R. W. Heckel, Met. Trans. *1*, 9 (1970).
[41] M. Hillert, Yernkontorets Ann. *141*, 67 (1957).
[42] R. A. Oriani, Acta Met. *12*, 1399 (1964).
[43] R. A. Oriani, Acta Met. *14*, 84 (1966).
[44] A. L. Hurley and A. M. Dayananda, Met. Trans. *1*, 139 (1970).
[45] J. Cahn, Trans. AIME *242*, 168 (1968).

8. Reactions between solids and gases or between solids and liquids with a solid reaction product

8.1. Oxidation of metals (tarnishing)

A tarnishing process is a reaction between a metal and a gas in which a solid product MeX_ν is formed according to the equation $Me(s) + \frac{\nu}{2} X_2(g) = MeX_\nu(s)$. X_2 could be oxygen, sulphur, chlorine, bromine, etc. An effective partial pressure of the electronegative component, for example of oxygen, can also be produced by auxiliary gas mixtures such as CO/CO_2 because of the reaction $CO_2 = 1/2\, O_2 + CO$. As long as this partial pressure of oxygen, which is determined by the ratio of partial pressures of CO and CO_2, is greater than the partial pressure of oxygen when the metal and metal oxide are in equilibrium, then the metal will be oxidized by the gas mixture. It is entirely proper that tarnishing processes should be included in a systematic study of solid state reactions, since the reaction product is a solid, and the reaction proceeds by the same elementary nucleation and growth steps as does a classical solid state reaction [9]. Transport will occur across the phase boundaries and through the phases. The rates of these transport processes will be determined by the chemical driving forces, and by the appropriate transport coefficients. The latter will depend upon the lattice defects and, in certain cases, upon the microstructure of the reaction product, since, for instance, gas could enter through cracks or pores in the oxidation product layer [10]. Furthermore, it is often possible to apply what has been learned from tarnishing experiments to the study of reactions between two solid phases. A great deal of data has been collected on tarnishing processes, as one might expect for a field of such technological importance. It is usually easier to adhere to simple experimental conditions in tarnishing processes than in reactions between solids. For instance, it is always difficult to maintain proper contact between solid phases, but there is no such problem at a gas/solid interface. Thus, it is not surprising that reaction rate laws have been determined accurately and unequivocally with simple experimental conditions for many more tarnishing reactions than for solid state reactions. A substantial fraction of the experimental data which has been amassed for tarnishing reactions has been interpreted atomistically, and the rate-determining elementary steps for the reactions have been definitely identified. Of course, if the product layer is very thin, or if it is cracked or not completely adherent, then the rate law for the oxidation may no longer be simple and easy to interpret. Similarly, the oxidation of an alloy may also be quite complex. The empirical rate laws for such cases may be of much practical interest, but they are less useful for providing an insight into the basic reaction mechanism.

In accordance with the expressed purpose of this monograph, we shall now discuss various models to explain the chemical and physical processes of tarnishing, and we shall not be concerned with presenting the details for particular systems. Such information can be found in a large number of special monographs on this subject [1, 2, 3].

In the year 1919, Tammann performed the first quantitative tarnishing experiments, using Ag in an iodine atmosphere. He observed a parabolic rate law for the increase in thickness of the product layer which he measured directly by observing the interference contours of visible light [48]. Three years later, an independent set of experiments were carried out by Pilling and Bedworth who measured the increase in weight of copper and other metals during oxidation in air. They also observed parabolic rate laws [11]. In the following years, a variety of different systems were studied, particularly by gravimetric, manometric, and volumetric techniques. As well as parabolic rate laws, many other rate equations were observed.

The particular rate law which will be obeyed by the reaction

$$\mathrm{Me(s)} + \frac{\nu}{2}\,\mathrm{X_2(g)} = \mathrm{MeX_\nu(s)}$$

will depend upon the type of metal Me, upon the time period of the reaction, and upon the experimental parameters such as temperature, partial pressure, sample shape, and others. For a one-dimensional experimental geometry, essentially the following laws have been observed to date:

$$\Delta m_{x_2} \propto \log t \tag{8-1}$$

$$\Delta m_{x_2}^{-1} \propto const - \log t \tag{8-2}$$

$$\Delta m_{x_2} \propto t \tag{8-3}$$

$$\Delta m_{x_2} \propto t^{1/2} \tag{8-4}$$

$$\Delta m_{x_2} \propto t^{1/3} \tag{8-5}$$

where Δm_{x_2} is the mass of gas X_2 which has been consumed. These rate laws are called, respectively: logarithmic, inverse logarithmic, linear, parabolic, and cubic. A theory of tarnishing reactions must explain these time-dependencies, and must show how the factors of proportionality (the rate constants) can be expressed in terms of simple physio-chemical quantities. It must be understood that these equations only describe limiting cases. If two or more elementary processes with similar reaction rates are superimposed, then a simple rate equation can no longer be expected. In the following sections, the various rate laws will be discussed and interpreted individually.

8.1.1. Parabolic rate law for the oxidation of metals

The parabolic rate law, which is named after Tammann [48] and Pilling and Bedworth [11], is by far the most important for heterogeneous reactions of the type $\mathrm{Me(s)} + \frac{\nu}{2}\,\mathrm{X_2(g)} = \mathrm{MeX_\nu(s)}$. Such a time-dependence is especially common for high-temperature corrosion. This rate equation can be interpreted on the basis of Wagner's theory of tarnishing, which is founded upon the same assumptions as were used in chapter 6 to derive the parabolic rate law for diffusion-controlled classical solid state reactions. The most important assumption is that thermodynamic equilibrium is maintained throughout the entire process at the phase boundaries $\mathrm{Me/MeX_\nu}$ and $\mathrm{MeX_\nu/X_2(g)}$, as well as locally within the compact, strongly adherent product layer $\mathrm{MeX_\nu}$. The validity of these assumptions, particularly as regards compactness and adherency, can only be tested experimentally for each individual case. Some rules were originally given by Pilling and Bedworth. Essentially, these rules state that an oxide will be porous if its molar volume is less than that of the metal consumed. There is no sound basis for such rules, however.

Under the given assumptions, the average gradient of the chemical potential, which is the driving force for the diffusional currents in the reaction product, is inversely proportional to the product layer thickness Δx at any time. Furthermore, the instantaneous rate of increase in thickness $\mathrm{d}\Delta x/\mathrm{d}t$ is porportional to $1/\Delta x$. This relationship can be integrated to give the parabolic rate law for tarnishing reactions in the form:

$$\Delta x^2 = 2\,\bar{k}\,t \tag{8-6}$$

In order to calculate the practical tarnishing reaction rate constant (Tammann constant) $\bar{k}$, Wagner assumed that ions and electronic charge carriers are responsible for mass transport through the product layer. This reaction scheme is shown in Fig. 8-1. Local electroneutrality must always be observed, of course.

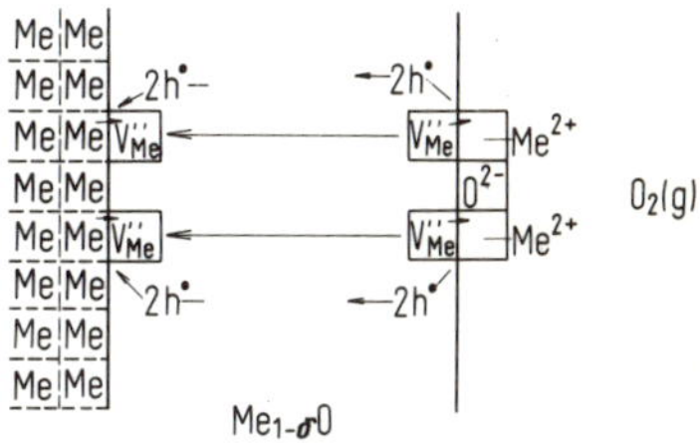

Fig. 8-1. Atomistic processes during the oxidation reaction $Me + \frac{1}{2}O_2(g) = MeO$.

The constant $\bar{k}$ can be calculated from the flux equations (5-13), with the condition of electroneutrality being used to eliminate the diffusion potential ϕ. The calculation is performed just as in the derivation of the rational rate constant for spinel formation in section 6.2.1. According to eq. (6-22), $\bar{k} = k\,v$, where v is the increase in volume of the product layer following the passage of one ionic equivalent. k is the rational tarnishing rate constant as introduced by Wagner [12]. It is equal to the flux in equivalents per unit area per unit time for a unit product layer thickness. By the method outlined above, k may be calculated as:

$$k = \frac{1}{z_{Me} F^2} \int_{\mu_{Me}(O_2)}^{\mu^0_{Me}} t_{ion}\, t_{el}\, \sigma\, \mathrm{d}\mu_{Me} \tag{8-7}$$

where z_{Me} is the valence of the metal ions, t_{ion} and t_{el} are the respective transport numbers, and σ is the total electrical conductivity. The transport numbers and conductivity can be replaced by diffusion coefficients through the use of eq. (5-16) to give:

$$k = \frac{2}{RTV_{MeO}} \int_{\mu_{Me}(O_2)}^{\mu^0_{Me}} t_{el}(D_{Me} + D_O)\, \mathrm{d}\mu_{Me} \tag{8-7a}$$

Furthermore, for the tarnishing of divalent transition metals (to form NiO, CoO, FeO, etc.), the electronic transport number can be set equal to one. The practical rate constant is then:

$$\bar{k} = \frac{1}{RT} \int_{\mu_{Me}(O_2)}^{\mu^0_{Me}} (D_{Me} + D_O)\, \mathrm{d}\mu_{Me} \tag{8-8}$$

The integration extends from the Me/MeO phase boundary where the chemical potential of the metal is μ_{Me}°, to the phase boundary $MeO/O_2(g)$, where the chemical potential of the metal is $\mu_{Me}(O_2)$. If the Gibbs free energy of formation of MeO, ΔG_{MeO}^{0}, is known, then this latter potential can easily be calculated, because of the equilibrium condition for $Me + 1/2\,O_2 = MeO$ at the $MeO/O_2(g)$ phase boundary. Average diffusion coefficients can be defined as:

$$\bar{D}_i = \frac{1}{\mu_{Me}^{\circ} - \mu_{Me}(O_2)} \int_{\mu_{Me}(O_2)}^{\mu_{Me}^{\circ}} D_i \, d\mu_{Me} \tag{8-9}$$

That is, the averages are taken with respect to chemical potential across the product layer. Furthermore, for transition metal oxides, the anion diffusion can be neglected relative to the cation diffusion ($D_O \ll D_{Me}$). Eq. (8-8) can thus be simplified to:

$$\bar{k} = \bar{D}_{Me} \frac{|\Delta G_{MeO}|}{RT} \tag{8-8 a}$$

where ΔG_{MeO} is the Gibbs free energy of formation of MeO from Me and from oxygen gas at the partial pressure of the particular experiment. The physico-chemical interpretation of the practical parabolic tarnishing rate constant is clearly evident in eq. (8-8 a). Corresponding to its dimensions of cm^2/sec, $\bar{k}$ varies as the product of the average component diffusion coefficient of the rate-determining partner and the driving force for the reaction expressed as the Gibbs free energy of formation ΔG_{MeO}. Of course, ΔG_{MeO} only appears as a ratio with RT, in accordance with the fact that the drift of the particles in the chemical potential gradient is only a fraction of their temperature-induced random motion. Naturally, the component diffusion coefficients can be replaced by other measures of the particle mobilities by the use of eqs. (5-14) or (5-16). Also, it is not difficult to modify the calculations for other stoichiometries (Al_2O_3), or for cases in which the product layer exhibits predominantly ionic rather than electronic conductivity, as in the case of bromides and chlorides. The details of such calculations can be found in the monographs which have been cited. For example, for products with predominantly ionic conductivity such as silver or copper halides, the reaction constant is given directly from eq. (8-7) as:

$$\bar{k} = \frac{\overline{\sigma_{el}}\, V_{MeX}}{F^2} \Delta G_{MeX} \tag{8-10}$$

where $\overline{\sigma_{el}}$ is the average electronic partial conductivity over the product layer, and F is the Faraday constant.

If the practical parabolic rate constant has been experimentally determined according to eq. (8-6) from measurements of the thickness of the product layer, then it will frequently be desirable to calculate the weight increase as a function of the product layer thickness. The corresponding rate constant for the weight increase has been named after Pilling and Bedworth. If q is the area of the metal, and ϱ_{MeO_v} is the density of MeO_v, then:

$$\Delta m = \Delta x\, q\, \varrho_{MeO_v} \frac{16\, v}{M_{MeO_v}} \tag{8-11}$$

where M_{MeO_v} is the molecular weight of the reaction product, and 16 is the atomic weight of oxygen. The practical Tammann rate constant is defined as $\bar{k}_T = \Delta x^2/2t$. The Pilling and Bedworth constant is defined as $\bar{k}_{PB} = \Delta m^2/q^2 t$. The two constants are thus related by the following equation:

$$\bar{k}_{PB} = \bar{k}_T \, 2\varrho^2_{MeO_\nu} \frac{16^2 \, \nu^2}{M^2_{MeO_\nu}} \tag{8-12}$$

If the range of homogeneity of the reaction product is sufficiently narrow, then the average diffusion coefficient as defined in eq. (8-9) can be calculated by means of defect thermodynamics, if it is assumed that the defects behave as the solute in ideally dilute solutions. In section 4.2 it was shown how the concentrations of the defect centers depend upon the component activities for a given type of disorder in binary ionic crystals. As an example, let us consider the formation of copper(I)oxide on copper sheet at 1000 °C in an oxidizing atmosphere when p_{O_2} is about 1 torr. The following defect equilibrium can be written:

$$\tfrac{1}{2}O_2 + 4\,Cu_{Cu} = 2\,V'_{Cu} + 2\,h^{\bullet}\,(Cu^{2+}_{Cu}) + Cu_2O \tag{8-13}$$

For the disorder type in Cu_2O: $(h^{\bullet}) \simeq (V'_{Cu})$. Thus, the following equilibrium condition holds for reaction (8-13):

$$(V'_{Cu}) = K_{13}^{1/4} \, p_{O_2}^{1/8} \tag{8-14}$$

That is, the concentration of cation vacancies is proportional to the eighth root of the partial pressure of oxygen in the case of the above mentioned ideal conditions. Now, in Cu_2O the transport number of the electronic charge carriers is one, and the diffusion of copper ions via vacancies is rate-determining. Thus, if local defect equilibrium is assumed, it follows from eq. (8-14) that the component diffusion coefficient D_{Cu} varies as $p_{O_2}^{1/8}$ according to the equation:

$$D_{Cu} = D^{\circ}_{Cu} \left(\frac{p_{O_2}}{p^{\circ}_{O_2}}\right)^{1/8} \tag{8-15}$$

where the superscript $^{\circ}$ denotes the equilibrium value at the Cu/Cu_2O phase boundary. Because of the overall equation $2\,Cu + 1/2\,O_2 = Cu_2O$, the following condition is fulfilled at all times at every point within the reaction product provided that local equilibrium is maintained:

$$2\mu_{Cu} + \tfrac{1}{2}\mu_{O_2} = \mu^{\circ}_{Cu_2O} \tag{8-16}$$

From the definition:

$$\mu_i = \mu_i^{\circ} + RT \ln \frac{p_i}{p_i^{\circ}}$$

and from eqs. (8-9), (8-15), and (8-16), the following expression for the average diffusion coefficient of the rate-determining partner can then be derived:

$$\bar{D}_{Cu} = \frac{1}{\Delta\mu_{Cu}} \int_{\mu_{Cu}(O_2)}^{\mu^{\circ}_{Cu}} D_{Cu} \, d\mu_{Cu} = \frac{4RT}{\Delta G_{Cu_2O}} D^{\circ}_{Cu} \left[1 - \left(\frac{p_{O_2}}{p^{\circ}_{O_2}}\right)^{1/8}\right] \tag{8-17}$$

where p_{O_2} is the oxygen partial pressure at the Cu_2O/O_2 phase boundary. If it is assumed that $(p_{O_2}/p^{\circ}_{O_2})^{1/8} \gg 1$, then the following expression for the rational reaction rate constant is obtained through substitution of eq. (8-17) into eq. (8-8a):

$$\bar{k} \approx 4 D^{\circ}_{Cu} \left(\frac{p_{O_2}}{p^{\circ}_{O_2}}\right)^{1/8} \approx 4 D_{Cu} \{ \text{in } Cu_2O\,(+O_2)\} \tag{8-18}$$

As can be seen, $\bar{k}$ is a function of the activity-dependent diffusion coefficient of copper in Cu_2O, and, most importantly, $\bar{k}$ is a function of the partial pressure of the oxygen gas.

In the above derivation of the rate constant for the parabolic growth of Cu_2O, all of the important aspects of a general calculation involving this type of tarnishing process have been discussed. For the case of a *n*-type conductor (e.g. ZnO on Zn), an analogous calculation would show that, in the first approximation, the practical rate constant $\bar{k}$ is independent of the partial pressure of oxygen. It can be seen immediately from the defect concentration gradients shown in Fig. 8-2 that this should be so, since the absolute change in defect concentration with oxygen potential at the oxide/gas interface is much smaller in case of a *n*-type material as compared to a *p*-type material.

The calculation of the tarnishing rate constant can become very complicated if different types of disorder occur within the range of component activities across the reaction product (e.g. in ZrO_2 [13]). Similarly, simple expressions for the rate constant cannot be expected if the reaction product possesses a wide range of homogeneity such that the defects no longer obey the laws of ideal dilute solutions, or if a multiphase product layer is formed. In the latter case, it can be shown, in analogy to the solid state reaction problem in section 6.2.3, that a parabolic growth law will still be obeyed for the total product layer thickness. However, the parabolic growth rate constant of each individual phase in the product layer will depend upon the thermodynamic and kinetic parameters of all the other phases [14, 15]. This problem has been discussed extensively in section 7.2 in connection with the definition of reaction rate constants of the first and second kind, which can be applied in the present case of multilayer tarnishing reaction products as well. An explicit expression for the ratio of the thickness of the two product phases in terms of the reaction volumes and the rational reaction rate constants has been given in reference [58]. If the free energies of formation of the individual phases in a multiphase product layer are approximately equal, but if the transport coefficients in the phases are very different because of differences in structure, then the overall thickness of the product layer may be determined essentially by only one of the product phases. The same is true if the chemical potentials of the metalloid components at the two interfaces of the outer product phase layer are approximately equal. (An example is $Cu/Cu_2O/CuO/O_2$ [16].) If the thinner phase of the product layer is in contact with the gas phase, as is the case in the oxidation of copper, then the overall rate constant will be almost independent of the partial pressure of the oxidizing gas, since the overall product layer thickness is essentially determined by the thickness of the Cu_2O phase, and the driving force for the growth of this phase is fixed by the invariant conditions at the two phase boundaries (Cu/Cu_2O and Cu_2O/CuO).

The essential features of the Wagner theory of tarnishing may be summarized with reference to eqs. (8-7) or (8-8). The hypothesis that transport in the product layer occurs by means of charged particles requires that there be a simultaneous diffusion of ions and electronic charge carriers, since otherwise electroneutrality would not be maintained. The overall rate

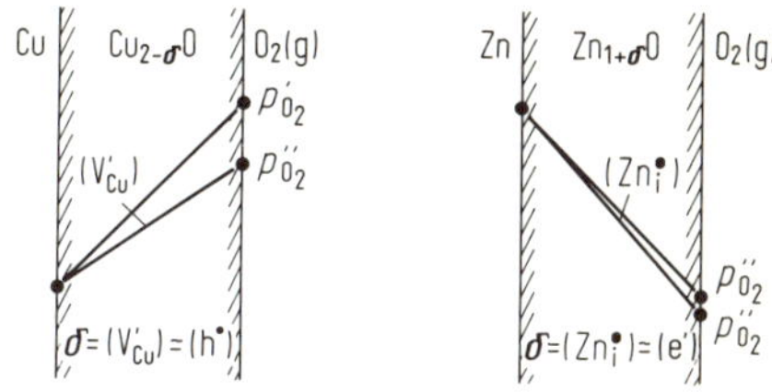

Fig. 8-2. Defect concentration gradients for p-conducting ($Cu_{2-\delta}O$) and n-conducting ($Zn_{1+\delta}O$) oxidation product layers for different pressures of the oxidizing gas ($p'_{O_2} > p''_{O_2}$).

will then be controlled by the partner which possesses the smaller transport coefficient (i.e. the smaller partial conductivity). Thus, if a material is to be oxidation-resistant, the oxidation product should have a negligibly small electronic conductivity (Al_2O_3) or, failing this, the ionic conductivity should be as low as possible. The defect thermodynamics discussed in section 4.2 provides guide lines as to which heterovalent dopants can best be used to achieve this goal. For example, if NiO, with cation vacancies as majority defects, is doped with Li_2O, lithium ions substitute for nickel ions. Thus, they form singly charged negative defects and decrease the number of cation vacancies and the cation transport number accordingly.

To conclude this section, a useful analogy between the tarnishing process in the present case and a working galvanic cell will be discussed. This presentation was first made by Jost [17] (see Fig. 8-3).

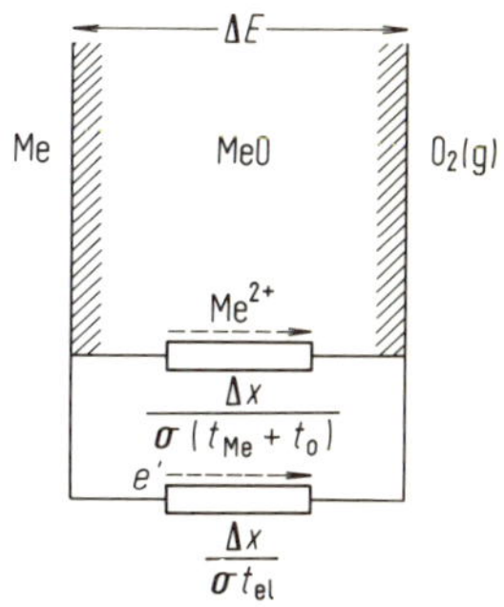

Fig. 8-3. Equivalent electrical circuit for tarnishing processes according to Jost [17].

The electrical work equivalent for ΔG_{MeO} is $z\Delta EF$, where z denotes the valence of the ions, and F is the Faraday constant (see also section 9.2.1). The electrical potential difference ΔE gives rise to an electronic current and an ionic current of equal magnitude with current density I through the product layer. According to Kirchhoff's law: $\Delta E + I(R_{el} + R_{ion}) = 0$. From Ohm's law, the resistances are given as:

$$R_{ion} = \frac{\Delta x}{\bar{\sigma}(\bar{t}_{Me} + \bar{t}_O)}, \quad R_{el} = \frac{\Delta x}{\bar{\sigma}\bar{t}_{el}}$$

$\bar{t}_i$ is the average transport number of species i, $\bar{\sigma}$ the average electrical conductivity.

Furthermore:

$$\frac{d\Delta x}{dt} = I\frac{V_{MeO}}{zF} \tag{8-19}$$

Making the appropriate substitutions, we obtain the following rate equation:

$$\frac{\mathrm{d}\Delta x}{\mathrm{d}t} = \frac{\Delta G_{\mathrm{MeO}}}{z^2 F^2} V_{\mathrm{m}} \bar{\sigma} \bar{t}_{\mathrm{el}} (\bar{t}_{\mathrm{Me}} + \bar{t}_{\mathrm{O}}) \frac{1}{\Delta x} \tag{8-20}$$

If this equation is combined with eqs. (8-6) and (5-16), it follows that:

$$\bar{k} = \frac{\Delta G_{\mathrm{MeO}}}{RT} \bar{t}_{\mathrm{el}} (\bar{D}_{\mathrm{Me}} + \bar{D}_{\mathrm{O}}) \tag{8-21}$$

in agreement with eqs. (8-8) and (8-9). The bars over the transport parameters once again indicate that these are average values taken over the reaction product layer [49].

8.1.2. Linear rate law for tarnishing processes

If the necessary conditions for a parabolic growth rate are not fulfilled, then other rate laws will be obeyed. One may find, for example, that the reaction rate is constant (i.e. independent of the instantaneous thickness of the product layer). However, this observation alone does not permit one to decide upon a definite rate-controlling elementary reaction mechanism, since several mechanisms can give rise to an overall linear rate law.

If the product layer is so thin that the diffusional resistance is negligible, and if the influence of electrical double layers at the phase boundaries (which will be discussed later) is also negligible, then only the phase boundary itself provides a barrier to mass transport. This means that thermodynamic equilibrium is no longer maintained at the phase boundary. Rather, there will be a discontinuity in activity here. The jump in activity is the driving force for transport across the phase boundary. The maximum value of this activity difference is determined by the free energy of formation of the reaction product. For a given value of the jump in activity, the resistance to transport across the phase boundary can be described with the aid of a transfer number which is formally analogous to the transport number in the Nernst boundary layer theory. If the diffusional resistance is negligible, then the jump in activity, which is the driving force for transfer across the phase boundary, is a constant, as is the transfer number itself. Consequently, the reaction rate is constant. That is, a linear rate law is obeyed. In general, the transfer number will depend upon the indices of the crystal plane at the interface. For instance, in a cubic system different transfer numbers would result when the (100), (110), or (111) planes were at the interface.

Certain individual steps during the adsorption or chemisorption of the reactant gas can also be rate-determining and can give rise to a linear rate law. In the simplest case, if diffusion in the reaction product and transport across the phase boundary are both sufficiently fast, then the supply of adsorbed gas molecules or gas atoms could become rate-controlling. This situation is most likely if the gas pressure and the adsorption coefficients are both low. Another example which belongs to the class of reactions discussed in this section is the oxidation of iron to wüstite in a CO_2/CO gas mixture when the thickness of the product layer is small ($T = 700\,°\mathrm{C}$, $\Delta x < 10^{-3}$ cm). The slowest elementary step is the dissociation of CO_2 into CO and adsorbed O atoms which are then rapidly incorporated into the wüstite lattice [18, 19].

The rate of this and many other such phase boundary reactions depends upon the instantaneous state of the surface. For tarnishing processes, this generally means that the rate depends upon the instantaneous activities of the components at the phase boundary as well as upon the temperature. As long as diffusional equilibrium is maintained, and the outer phase boundary reaction alone is rate-controlling, the activity of the metal at the phase boundary between oxidation product and gas is constant and equal to one. There are indications [50] that the electronic defects can particularly influence the rate of dissociation of the gases at the phase boundary between oxidation product and gas. Use is made of this property of solid surfaces in the field of heterogeneous catalysis [4]. Since the defect concentration is determined by the activities of the components in the reaction product, it is understandable that the rate of the phase boundary reaction should, in general, depend upon the component activities in the reaction product at the phase boundary.

8.1.3. Logarithmic rate law for tarnishing

In addition to the parabolic and linear growth laws, other rate equations for tarnishing processes have been observed, particularly in the early stages of these reactions (see eqs. (8-1), (8-2), and (8-5)). These can be attributed essentially to the influence of electrical space charges in the tarnishing product layers resulting from chemisorbed gas molecules at the surface, and to the effect of electrical double layers at the phase boundary between metal and oxidation product. The time-varying electrical fields connected with these effects will influence the rate of migration of the ions in the reaction product. Logarithmic rate laws are seldom observed during high temperature oxidation. However, they are frequently found for the formation of thin surface oxide layers on metals at lower temperatures. It is very difficult to theoretically interpret the results on the basis of a model, since it is usually not possible to independently measure the parameters of the model for the very small layer thicknesses of from 10 Å to about 100 or 1 000 Å with which we are concerned here.

The basic parameter in a model of such processes is the Debye-Hückel length $l = \left(\frac{\varepsilon k T}{8 \pi n_i e_0^2}\right)^{1/2}$ which was discussed in section 3.4. If the product layer thickness Δx is very small compared to l, then the layer can be compared to a condenser. The charge on the outer interface of the tarnishing product (surface charge) is provided by negatively-charged chemisorbed gases such as O_{ad}^{n-}, S_{ad}^{n-}, etc. Different logarithmic rate laws will result, depending upon which is the rate-determining step for the reaction of the metal cations with these electronegative chemisorbed gases. For example, a simple logarithmic rate law (eq. (8-1)) will result when, for a constant specific surface charge, the rate-determining step is the tunnelling of electrons out of the metal through the reaction layer to the solid/gas phase boundary. In this case, the X_{ad}^{n-} ions can react with metal ions and become incorporated into the product layer only as fast as electrons are supplied to form new X_{ad}^{n-} ions. Therefore, $\mathrm{d}\Delta x/\mathrm{d}t$ is proportional to the tunnelling probability: $\exp - \beta \Delta x$. Eq. (8-1) then follows directly by integration.

On the other hand, if the electronic conductivity of the product layer is high enough, ionic transport in the electrical field of the "condenser" may become rate-controlling. In this case, an inverse logarithmic rate law results, since the activation energy for the motion of an ion in the product layer decreases when there is a constant electrical potential difference across the layer. The amount by which the activation energy decreases is proportional to the electric field strength in the condenser, and the decrease is thus inversely proportional to the thickness of the product layer. Consequently, the ionic flux is proportional to $\exp -\frac{E_A - \text{const}/\Delta x}{RT}$, where E_A is the activation energy in the absence of an electric field. Since $j_i \propto \mathrm{d}\Delta x/\mathrm{d}t$, the inverse logarithmic rate follows directly by integration, as long as the product layer is not too thin. The details of the calculations can be found in the special literature [20, 21].

If the thickness of the product layer is of the order of the Debye-Hückel length, then the electric field which affects the transport of particles in the product layer will be modified by space charges. This will influence the overall reaction rate for the case in which transport in the reaction product layer is rate-determining. Furthermore, when Δx is of the order of the Debye-Hückel length, the density of charge carriers at the surface will depend upon the thickness of the product layer, and so the overall reaction rate will also be affected in the case when adsorption processes are rate-controlling, as was discussed in the previous section. For the first case, Hauffe [21] has shown that a cubic rate law is indicated. However, such an interpretation of a cubic rate law is by no means unambiguous, since the dissociation of gas molecules during chemisorption, if rate-determining, can likewise result in a cubic rate law. In order to make a quantitative calculation, we assume that, in the simplest case, the rate of dissociation is a linear function of the concentration of electronic charge carriers at the surface, The latter statement has been verified several times [50]. We then calculate the variation of this concentration with the thickness of the product layer on the basis of the electronic boundary layer theory [5]. If we assume that local equilibrium is maintained at the Me/MeX_v phase boundary, then the concentration of electronic charge carriers at this phase boundary c_e° is fixed. The Boltzmann distribution of charge carriers in the film may now be written as:

$$c_e = c_e^\circ \exp \frac{F\phi}{RT} \tag{8-22}$$

We also require Poisson's equation which relates the electrical potential ϕ in the product layer to the dielectric constant ε and the space charge density ϱ:

$$\frac{\mathrm{d}^2\phi}{\mathrm{d}x^2} = -\frac{4\pi\varrho}{\varepsilon} \tag{8-23}$$

By integrating eq. (8-23) and using eq. (8-22), we can calculate the concentration of electronic charge carriers at the surface, c_e, and thus we can calculate the rate of dissociation, if we assume a certain type of disorder in the thin film layer, and if the proper boundary conditions are used. As an example [22, 23], let us consider the oxidation of Ni to NiO. For thin oxide layers, the oxidation process approximately obeys a cubic rate law between 250 °C and 400 °C. Simultaneous studies of oxidation and homomolecular isotope exchange according to the reaction $O^{16}O^{16} + O^{18}O^{18} = 2\,O^{16}O^{18}$ at the NiO surface have been carried out. These studies show that, under these conditions, the rate-determining step for the growth of the NiO layer is the dissociation of O_2 molecules on the surface. By means of the calculations outlined above, it can be shown that the electron concentration at the NiO surface varies inversely as Δx^2. Therefore, if it is assumed that the rate of dissociation is proportional to the concentration of electrons at the surface, a cubic rate law is obtained, in agreement with the experimental observations.

8.1.4. Tarnishing with simultaneous dissolution of gas in the metal

In this section, an important special case of high-temperature oxidation will be discussed. Up to now we have assumed that there is no appreciable dissolution of the electronegative component X in the metal during the oxidation process. However, in many important practical systems this is by no means the case, as one can easily appreciate by looking for example at the phase diagram for the Zr-O system. What rate law should we then expect for the growth of the product layer when dissolution of the element X in the metal occurs simultaneously with the growth of the oxidation product, if local equilibrium can be assumed? The situation for a one-dimensional experiment when the compound MeX is formed is illustrated in Fig. 8-4.

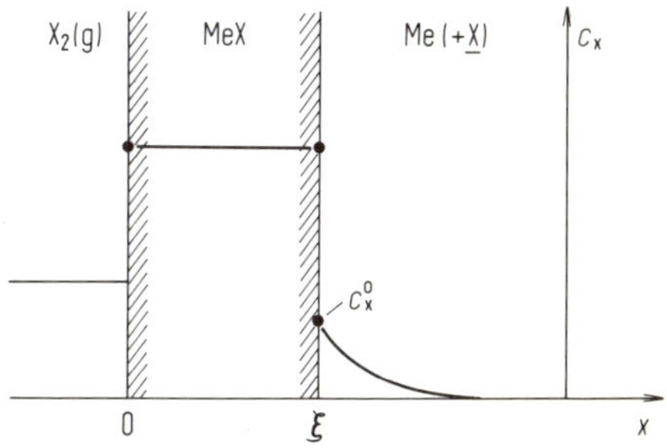

Fig. 8-4. Dissolution of gas in the metal during oxidation: coordinate system and concentration profile.

Analogous situations for the case of classical solid state reactions have already been mentioned in sections 6.2 and 7.2. An example is the reaction between CoO and Cr_2O_3 to form $CoCr_2O_4$ with the simultaneous dissolution of Cr_2O_3 in CoO. However, the dissolution of gases in metals during oxidation is a technologically important and, at the same time, a very illustrative example, and so the quantitative treatment of this problem will be outlined here. An important practical example is the oxidation of zirconium.

A very general solution of this problem and of the analogous problem for classical solid state reactions has recently been presented [57]. The following is a condensed treatment in which certain simplifying assumptions have been made. If local equilibrium is assumed, then the ionic flux j_i in the product layer MeX ($Me^{2+}X^{2-}$) is given as $j_i = k/2\,\xi$, where k is the rational tarnishing rate constant, as in eq. (8-7), and where ξ is the phase boundary coordinate in Fig. 8-4. ξ is equal to the instantaneous product layer thickness Δx. However, not all of the flux j_i contributes to the growth of the product layer, since a certain amount of X diffuses away into the metal. Since only diffusional processes are involved, the overall process will still obey a parabolic rate equation. Consequently, $\xi = \alpha\sqrt{t}$, where α is a rate constant which must now be calculated. In mathematical language we can write:

$$\frac{d\xi}{dt} = \frac{\alpha}{2\sqrt{t}} = V_{MeX}\left[j_i + \tilde{D}_X^{Me}\left(\frac{\partial c_X}{\partial x}\right)_\xi\right] = V_{MeX}\left[\frac{k}{2\xi} + \tilde{D}_X^{Me}\left(\frac{\partial c_X}{\partial x}\right)_\xi\right] \qquad (8\text{-}24)$$

The second term in square brackets accounts for the transport of X into the metal according to Fick's first law. The simplifying assumption has been made that c_X°, the equilibrium solubility of X in the metal, is small, so that the mole fraction of metal in the metal phase at $x = \xi$

is approximately equal to unity. If this cannot be assumed, then continuity equations must be written for X and for Me at $x = \xi$. The general problem can become very complicated. Also, the diffusion mechanism in both the metal and the MeX phase must now be known in order that the continuity equations can be properly formulated [57]. However, this is not the case in the simplified treatment presented here.

Continuing with the solution of (8-24), we must now solve Fick's second law for the diffusion of X into the metal with the boundary conditions $c_X \to 0$ for $x \to \infty$, and $c_X = c_X^\circ$ for $x = \xi$. According to eq. (5-42), this solution is:

$$c_X = B\left(1 - \operatorname{erf}\frac{x}{2\sqrt{\tilde{D}_X^{Me}t}}\right); \quad B = c_X^\circ\left(1 - \operatorname{erf}\frac{\alpha}{2\sqrt{\tilde{D}_X^{Me}}}\right)^{-1} \tag{8-25}$$

By taking the derivative $(\partial c_X/\partial x)$ at $x = \xi$ from eq. (8-25), and substituting this into eq. (8-24), we obtain the following transcendental equation for the growth rate constant α:

$$\alpha^2 + \frac{2\, V_{MeX} c_X^\circ \sqrt{\tilde{D}_X^{Me}}}{\pi^{1/2}} \cdot \frac{\alpha}{1 - \operatorname{erf}\alpha/2\sqrt{\tilde{D}_X^{Me}}} \cdot e^{-\frac{\alpha^2}{4\tilde{D}_X^{Me}}} = k\, V_{MeX} \tag{8-26}$$

If the transport parameters k and $\tilde{D}_X^{Me}$ and the molar volume V_{MeX} are known, then this equation can be solved graphically. It follows immediately from eq. (8-26) that, if the solubility of X in Me is negligible (i.e. if $c_X^\circ \to 0$), or if the mobility of X in Me is negligible ($\tilde{D}_X^{Me} \to 0$), then $\alpha^2 = k\, V_{MeX}$, and $\xi^2 = k\, V_{MeX}\, t = 2\,\bar{k}t$. Since $\xi = \Delta x$, eq. (8-26) reduces to the parabolic equation (8-6) in this limiting case.

With this final example, all the important fundamental aspects necessary for an understanding of tarnishing processes have now been discussed.

8.1.5. Oxidation of alloys

Alloys are by far the most important metallic materials from a technological standpoint. Thus, there is great interest in the field of alloy oxidation. The oxidation characteristics of alloys are a very important criterion for many practical applications, such as turbine blades, hot air motors, jet propulsion systems, etc. [6]. In contrast to the oxidation of pure metals, alloy oxidation processes exhibit certain unique aspects. For the sake of simplicity, we shall confine our discussion to binary alloys (A, B). The basic parameters in alloy oxidation are the composition of the alloy (measured in terms of mole fractions $N_A = 1 - N_B$), the number and the type of compounds and solid solutions in the ternary phase diagram A-B-O, the free energies of formation of these phases, and finally, the mobilities of all three components in the metal phase and in the product phases. Depending upon which of these parameters has the major influence upon the overall oxidation process, certain limiting cases of alloy oxidation can be distinguished [24]. These limiting cases can be treated quantitatively. However, the problem of alloy oxidation in its most general formulation is virtually insoluble because of the mathematical difficulties involved. Some important limiting cases will now be discussed.

1. Only the oxides AO and BO are formed, and $|\Delta G_{AO}^0| \gg |\Delta G_{BO}^0|$. That is, B is more noble than A. Furthermore, the metals A and B exhibit complete mutual solubility. As long as the

ratio of the mole fractions and the ratio of the mobilities in the alloy are not extreme, it can be assumed that the more stable AO is the only oxidation product which is formed. Therefore, there is an increase in the concentration of B in the alloy at the phase boundary, and a concentration gradient is built up in the alloy phase. This situation is illustrated in Fig. 8-5 for the oxidation of a Ni-Pt alloy [25, 26].

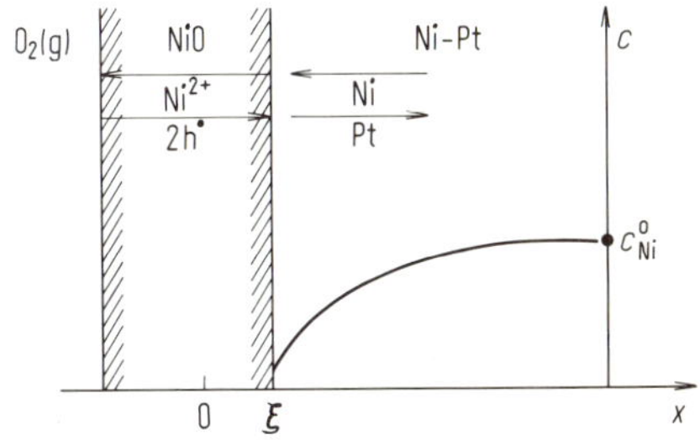

Fig. 8-5. Oxidation of a Ni-Pt alloy: concentration profile and coordinate system.

If local equilibrium is assumed, then the reaction kinetics can be calculated from the flux equation for A ions in the product layer and from the solution of Fick's second law in the alloy. The continuity condition at the phase boundary must also be observed. The calculations are analogous to those in sections 7.2.1 and 8.1.4. However, one additional important point must be considered. If diffusion in the alloy is the rate-controlling step in the overall process, then a slight disturbance in the planar metal/oxide phase boundary will be unstable. That is, if the phase boundary bows into the metal phase at some spot, then the growth rate of AO is increased at this point, and the disturbance increases in magnitude. The result is a fissured phase boundary. The morphology of this phase boundary depends upon a variety of factors such as the ratio of molar volumes V_{AO}/V_{alloy}, the plastic behaviour of the oxide and the metal, and the adherency between oxide and alloy. Examples of reactions in which strongly fissured phase boundaries form are the reactions of Ag-Au and Cu-Au alloys with sulphur at 400 °C [31].

If AO and BO are mutually soluble, then the situation will be different, especially if the stabilities of the two oxides are comparable. For example, a (Co, Ni) O product layer is formed when Co-Ni alloys are oxidized [51]. Since Co is preferentially oxidized (because it is less noble), the Co content of the alloy decreases. Also, the mole fraction of CoO in the product phase increases in the direction from the metal to the gas. This indicates that the mobility of Co ions in the oxide solid solution is greater than that of the Ni ions. The quantitative treatment of the growth of this quasi-binary reaction product has been given only in an approximate form using a number of simplifications because of mathematical difficulties. However, in some recent studies the application of this approximate theory has been shown to describe the tarnishing of (Co, Ni) or (Mn, Fe) alloys satisfactorily [59].

2. If the diffusion of a less noble metal A which is dissolved in a more noble metal B is relatively slow compared to the diffusion of dissolved oxidizing gas in the alloy, then internal oxidation may take place. This phenomenon, which has been observed and studied for many alloy systems, is particularly likely to occur when the oxidizing gas is very soluble in the metal, and when the concentration of the metal A is very low. The process is illustrated schematically in Fig. 8-6. Such a situation is observed, for example, during the oxidation of Ag alloys with small additions of Cu, Cd, or Al [27, 28].

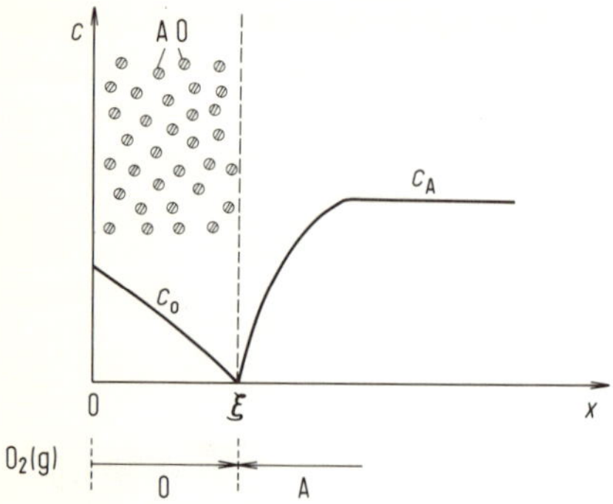

Fig. 8-6. Diffusional processes during internal oxidation: concentration profile and coordinate system. The oxide AO has precipitated up to the reaction front at $x = \xi$.

The oxidation of Ag-In alloys with small In contents proceeds in the same way. Oxygen dissolves in the silver alloy essentially as atomic oxygen, and then diffuses into the interior where it reacts to form small particles of In_2O_3. If the value of ΔG^0_{AO} (or $\Delta G^0_{A_2O_3}$, etc.) for the formation of the less noble metal oxide is sufficiently negative, then the concentration of A will be negligibly small in the region $x < \xi$ where there is an excess of oxygen. As long as diffusional processes are rate-controlling, the reaction front will advance into the metal according to a parabolic rate law. That is, $\xi = 2\,\alpha\sqrt{D_O t}$. The constant α may be calculated as a function of D_O, D_A, $N_O\,(x = 0)$, and $N_A\,(x = \infty)$ by means of Fick's second law for the diffusion of oxygen in the region $x < \xi$, and for the diffusion of A in the region $x > \xi$. It is also necessary here to make use of the condition of continuity at $x = \xi$, which says that the fluxes of oxygen and of metal A to the point $x = \xi$, where AO is being formed, must be equal. The following equation for α is obtained in a way which is analogous to the derivation of eq. (8-26):

$$N_O\,(x = 0)\sqrt{D_O/D_A}\cdot e^{\alpha^2 D_O/D_A}\cdot\left(1 - \operatorname{erf}\alpha\sqrt{D_O/D_A}\right) = N_A\,(x = \infty)\,e^{\alpha^2}\operatorname{erf}\alpha \qquad (8\text{-}27)$$

The condition for internal oxidation to occur can be written as $D_O N_O \gg D_A N_A$. That is, the transport coefficient of oxygen is much greater than that of the metal A. Eq. (8-27) can thus be simplified, and the following equation for α results:

$$\alpha \approx \left[\frac{N_O\,(x = 0)}{2\,N_A\,(x = \infty)}\right]^{\frac{1}{2}} \qquad (8\text{-}28)$$

Since there is virtually no transport of A in this case, there will be a uniform distribution of the precipitate AO in the region $x < \xi$, and the concentration of precipitated AO will correspond to the original mole fraction N_A of A in the alloy. Internal oxidation may nevertheless occur even if the transport coefficient of A is not negligible compared to the transport coefficient of oxygen. In this case, the concentration of precipitated AO in the zone of internal oxidation will become greater than the original concentration of A in this zone. With increasing enrichment of AO, the transition from internal to external oxidation will become more likely. This problem has been treated quantitatively by Wagner [24].

An important technological application of internal oxidation is in dispersion hardening. The fine particles of oxide can cause a considerable hardening of the matrix metal. A review of this subject is given in reference [52].

3. Under point 1., we considered the case in which only the oxide AO is formed during the course of alloy oxidation. As shown in Fig. 8-5, this results in a continual increase in the concentration of B in the alloy at the phase boundary. As long as the ratio of the free energies of formation of AO and BO is not too great, there will come a time when the concentration and the activity of B has increased to such an extent that BO will be nucleated at the alloy/oxide phase boundary. This will once again decrease the concentration of B in the alloy. Then, if B cannot diffuse out of the alloy to the phase boundary fast enough, this particle of BO will detach itself from the phase boundary, and will become completely surrounded by AO as shown in Fig. 8-7.

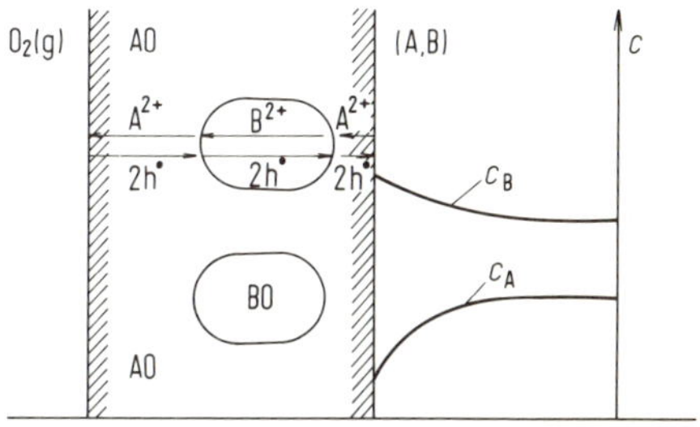

Fig. 8-7. Hypothetical reaction mechanism of alloy oxidation when a two-phase oxide layer is formed.

The phase boundary reactions on the right- and left-hand-sides of the BO particle in Fig. 8-7 are, respectively:

$$A^{2+}(AO) + BO = B^{2+}(BO) + AO \quad \text{and} \quad B^{2+}(BO) + AO = A^{2+}(AO) + BO$$

At the same time, electrons or electron holes must diffuse in the oxides in order to maintain electroneutrality. This is also illustrated in Fig. 8-7.

The oxidation of alloys can become very complicated as a result of mutual solubility of AO and BO [29], as a result of the formation of double oxides [30], or as a result of all those effects connected with changes in molar volume and with the adherency of the reaction products. Therefore, this matter will not be pursued any further here. In practice, the production of oxidation-resistant alloys must still be based to a great extent upon empirical rules [6], although the theoretical ideas as outlined here are a necessary prerequisite in order to do systematic oxidation – resistant alloy development.

These few limiting cases of alloy oxidation by no means provide an exhaustive treatment of the subject. A survey of the morphological aspects of this class of reactions is given in reference [53]. The complex process of alloy oxidation cannot be understood without a coordinated study with respect to the kinetics, composition, structure, and morphology of the reacting system. The theoretical tools of defect thermodynamics and diffusion theory can help to show the way, but they are not sufficient of themselves to provide a complete understanding in most cases.

8.2. Corrosion

Corrosion is the attack of a gaseous or liquid phase upon a metal or other solid with a resultant destruction of the structure. The fundamental aspects of the problem of gaseous corrosion were discussed in the previous section. Since we are concerned here with solid state reactions, we shall discuss only those corrosion processes in liquid media (i.e. particularly in aqueous electrolytes) in which a solid reaction product in the form of a surface coating is formed. As a corrosive medium, an aqueous solution exhibits two properties which fundamentally distinguish it from a gas phase: 1. The aqueous solution exhibits electrolytic conductance; and 2. It can dissolve the reaction product.

Although there is no external current, anodic and cathodic processes can still occur at sites on the interface between solid and aqueous solution because of the electrolytic conductance of the corrosive medium. At electrochemical equilibrium, this leads to a definite jump in the electrical potential at the phase boundary. Kinetic barriers to certain partial reaction steps of the electrochemical process can cause the potential to be displaced from its equilibrium value. Thus, for example, instead of a dissolution of metal:

$$Me = Me^{n+} + ne^- \qquad (8\text{-}29)$$

a solid product layer might form, according to the overall reaction:

$$Me + nOH^- = MeO_{n/2} + n/2\, H_2O + ne^- \qquad (8\text{-}30)$$

If the product layer is nearly free of pores, then the anodic dissolution of metal will practically cease. The metal is then said to be "passivated". The thickness of the compact product layer will reach a stationary value. For oxide products which are essentially electronic conductors, this stationary thickness will be determined by the very low ionic conductivity in the oxide on the one hand, and by the rate of dissolution of the oxide in the electrolyte on the other. However, in many cases the oxide layers are porous, so that the electrolyte can continue to attack the metal, independently of the transport of ions and electrons in the oxide. From the above discussion it can be seen that corrosion reactions in aqueous ionic solutions in which a solid product layer is formed on a metal are among the most complicated of all heterogeneous solid state reactions. The reasons for this are the electrochemical nature of these reactions, the great number of possible elementary steps which can occur at the various phase boundaries, and electrical space charge phenomena which occur in the reaction product.

In the case of those solid state reactions which we discussed in the previous sections, ordinary phase diagrams can be used to provide information as to which product phases will be formed between the reactants if local thermodynamic equilibrium is everywhere maintained. For the case of electrochemical corrosion, such information can be obtained from potential-p_H diagrams which were originally introduced by Pourbaix [32, 33]. These diagrams, which are usually calculated for reactions at room temperature and atmospheric pressure, give the stability limits of the metal and its compounds relative to the ions in the aqueous solution. The stability limits are plotted as lines on a diagram in which the essential electrochemical parameters, namely the p_H and the electrical potential of the electrode relative to the standard hydrogen electrode, have been chosen as abscissa and ordinate. To determine these stability limits, one calculates the equilibrium electrode potentials of the ions in the solution for a given ionic activity as well as the equilibrium electrode potentials of the metal and its various solid reaction products as functions of p_H, setting both terms equal and then plots these lines on the diagram.

For example, in the system Fe/H_2O, let us consider the two electrode reactions:

$$Fe = Fe^{2+} + 2\,e^- ; \qquad \Delta G_1 \qquad (8\text{-}31)$$

and

$$3\,Fe + 12\,H_2O = Fe_3O_4 + 8\,H_3O^+ + 8\,e^- ; \quad \Delta G_2 \qquad (8\text{-}32)$$

If we multiply eq. (8-31) by -3 and add it to eq. (8-32), we can see that if $\Delta G_2 - 3\,\Delta G_1 < 0$, then a Fe_3O_4 surface layer will be stable. However, if $\Delta G_2 - 3\,\Delta G_1 > 0$, then iron will be in equilibrium with Fe^{2+} ions in aqueous solution. By choosing a suitable standard state, we can relate the free energies ΔG_i to the equilibrium electrode potentials E_H by the equation $\Delta G_i = -zFE_{H,i}$. Since the ΔG_i are functions of the chemical potentials μ_i of the metal ions and of the hydrogen ions, the equilibrium lines in the Pourbaix-diagram will be dependent upon p_H.

The stability limits of water are fixed by the following equilibria:

$$H_3O^+ + e^- = H_2O + \frac{1}{2}H_2 \tag{8-33}$$

$$2\,H_2O = H_3O^+ + OH^- \tag{8-34}$$

From the equilibrium conditions of these two reactions, the equilibrium potentials for the evolution of H_2 or O_2 from aqueous solutions can be calculated as a function of p_H. For $P =$ 1 atm:

$$\left.\begin{aligned} E_H(H_2) &= -\,0.059\,p_H \qquad &\text{(I)}\\ E_H(O_2) &= 1.23 - 0.059\,p_H \qquad &\text{(II)}\end{aligned}\right\} \tag{8-35}$$

In this way, then, it is possible to state the limits of those regions in which iron for example is stable, is passivated, or corrodes in aqueous solution.

It must be remembered, however, that the Pourbaix-diagrams are based solely upon thermodynamic calculations, and in reality, the stability limits can be displaced to a considerable extent as a result of kinetic restraints. The same comment, of course, can be made with regard to the application of ordinary phase diagrams in the interpretation of solid state reactions or tarnishing processes in which a multiphase product is formed, as has been pointed out, for example, in section 6.2.3.

8.2.1. Kinetics of formation and dissolution of compact corrosion products

The theory of the kinetics of formation and dissolution of compact corrosion products is based upon the same concepts as is the theory of the growth kinetics of thin tarnish coatings. The only difference in the physical picture of the two cases is that, in the case of aqueous corrosion, there will also be an electrical space charge region in the aqueous electrolyte. This, together with the chemisorbed ions on the surface of the solid corrosion product, will determine the eletrical space charge, and thus also the electric field, in the corrosion product. Once again, as in the case of tarnishing, the overall kinetics will depend upon which elementary step is rate-controlling. Possibilities are: the transfer of metal ions across the two phase boundaries, the transport of ions or electrons through the corrosion product, or the incorporation of oxygen ions from the chemisorbed layer or from the electrolyte into the corrosion product. The last mentioned possibility may involve a number of coupled reactions in the liquid electrolyte. A detailed discussion of the kinetics for each of these possibilities can be found in the special literature [7]. This discussion then is part of the large field of electrode kinetics in electrochemistry, and the interested reader is referred to its modern monographs.

Finally, we should briefly discuss the rate of steady state dissolution of a compact corrosion product in aqueous solution. In order to calculate this stationary corrosion rate, it is necessary to know the rate-determining step for the removal of metal ions from the corrosion product into the electrolytic solution. It has been found that the stationary corrosion rate of a passivated layer on iron between $p_H = 0.7$ and 3.9 in an aqueous H_2SO_4 solution does not depend upon the stirring rate, nor upon the Fe^{2+}/Fe^{3+} ratio in the solution, nor, most importantly, does it depend upon the electrode potential E_H [34]. From these observations it has been concluded that the rate-controlling step is a chemical reaction between the iron ions, which are leaving the corrosion product, and the solvent ions to form complex ions immediately next to the surface. Thus, in the case of iron under the given conditions, this step determines the overall corrosion rate and thus it also determines the stationary thickness of the passivating corrosion product.

From these few remarks regarding the formation and dissolution of compact corrosion products, it can be appreciated how complex these technologically important processes can be. Fundamentally different situations will be encountered from one system to the next. Schwenk [54] has recently reviewed the extensive studies which have been made of the corrosion of steel.

8.3. Topochemical reactions

Topochemical reactions in the sense of this section are solid state reactions under the action of a chemical potential gradient that occur essentially at distinguished sites of the reactant. There are a large number of classical solid state reactions as well as tarnishing reactions in which the morphology of the reaction product is a result of the existence of fast transport paths in which the transport coefficient of a reactant is relatively high. The morphologies can often be quite unusual and interesting. Examples of fast transport paths are surfaces, grain boundaries, pores, and the surrounding gas atmosphere. In such topochemical reactions, the rate of advancement of the reaction front is no longer controlled by the diffusion of the slower partner in the reaction product. Rather, it is determined by the diffusion of the faster partner, since the regions of fast transport ensure that the other partner will always be present at the reaction site. Topochemical reactions could have been treated systematically in chapter 6. However, many of the best examples of such processes are gas-solid reactions, and so the discussion has been postponed until now. An experiment which has already become a classic is shown in Fig. 8-8.

A tantalum substrate has been dipped in molten silver so that it is partially covered by a thin adherent layer of solid silver. Just as in Tammann's [48] original experiment, the silver is iodized in an atmosphere of iodine gas. The AgI surface layer grows according to a parabolic rate law. Since AgI is virtually a purely ionic conductor (i.e. $t_{el} \ll 1$), the product layer will be very thin as can be seen from eq. (8-7). It is also observed, however, that the AgI spreads out along the un-silvered portion of the tantalum foil at a relatively fast parabolic rate. For this sideways growth, the electrons need no longer diffuse through the entire width of the AgI film, since they can now move with practically no resistance through the metallic tantalum substrate to the reaction front. Therefore, the diffusion of the much faster silver ions in the AgI becomes the rate-determining step for sideways growth along the tantalum foil. From the ratio of the thickness of the AgI layer, Δx_1, to the extent of the sideways growth, Δx_2 (see Fig. 8-8), it is possible to calculate the average transport number of the electrons in AgI. From eqs. (8-6) and (8-8), it follows that:

$$\left(\frac{\Delta x_1}{\Delta x_2}\right) = \left(\frac{\bar{k}_1}{\bar{k}_2}\right)^{\frac{1}{2}} \approx \left(\bar{t}_{el}\frac{\bar{D}_{Ag}}{\bar{D}_{Ag}}\right)^{\frac{1}{2}} = \bar{t}_{el}^{\frac{1}{2}} \tag{8-36}$$

It is important to note here that the average values (designated by bars) are not arithmetic means of the values at the Ag/AgI and AgI/I_2 boundaries, but rather, they are averages with respect to chemical potential across the AgI layer as in eq. (8-9). The activity dependence of the transport coefficients is determined by the disorder type and by defect thermodynamics, as has been discussed many times in this monograph.

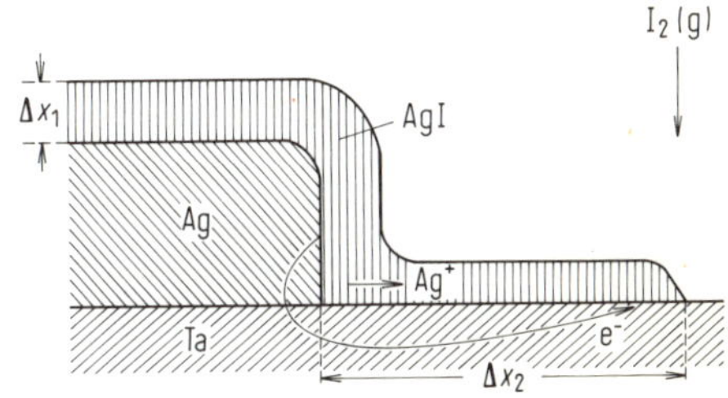

Fig. 8-8. The morphology of the reaction Ag $+ \frac{1}{2}I_2(g) = AgI$ upon a tantalum substrate [35].

A similar experiment has recently been performed to study the growth kinetics of AgBr on Ag in a bromine atmosphere [36]. A thin Pt-gauze, which was permeable to the gas, was affixed to the surface of the AgBr. The rate of growth of the AgBr layer was measured both when the Pt-gauze was externally connected to the Ag (short circuit), and when it was not (open circuit). The rate of growth after short-circuiting was much greater than the rate at open circuit, since in the former case the electrons are transported through the external circuit, and the rate-determining step is the relatively fast diffusion of the ions in the AgBr, whereas at open circuit, the rate-controlling step is the relatively slow diffusion of electronic charge carriers in the AgBr. The original data are given in Fig. 8-9. From the ratio of the growth rates for a given Δx at short circuit and at open circuit, the average transport number of the electronic charge carriers in the product layer can be calculated from eq. (8-36).

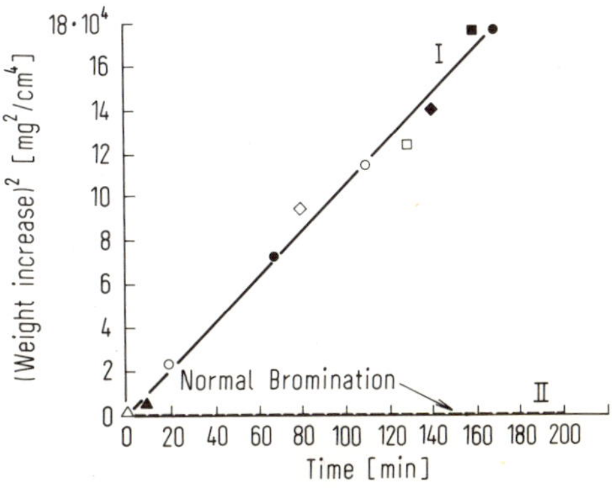

Fig. 8-9. Parabolic growth of a AgBr product layer on Ag in Br_2 gas ($T = 400$ °C, $p_{Br_2} = 175$ torr). (I) Short circuit between the Ag and a Pt-gauze on the AgBr surface. (II) Normal growth of a compact AgBr product [36].

There are many solid state reactions with interesting morphologies in which other modes of transport besides ionic transport in the reaction product can be very important. As a first example, let us consider a reaction which normally occurs by the counter-diffusion of cations, as illustrated in Fig. 2-1 for the formation of spinels in the reaction couple $AO/AB_2O_4/B_2O_3$. It was assumed in this case that the mobility of the oxygen ions is negligible. However, if the reactant oxides AO and B_2O_3 are very porous, then the reaction product AB_2O_4 will also be porous. Furthermore, if the reaction product is an electronic conductor ($t_{el} \approx 1$), as is generally the case when the ions A and/or B are transition metal cations, then oxygen can be transported via the gas phase through the pores, while diffusion of electrons (or electron holes) and ions occurs simultaneously in the reaction product. This situation is illustrated in Fig. 8-10.

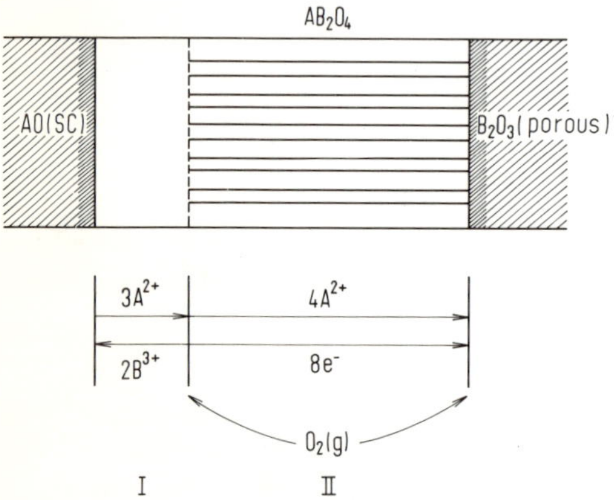

Fig. 8-10. Reaction mechanism for the formation of AB_2O_4 from a single crystal AO(SC) and a porous oxide B_2O_3 (porous) by the overall reaction $AO + B_2O_3 = AB_2O_4$. (I) Region where the spinel is dense, and reaction occurs by counter-diffusion of cations. (II) Region of porous spinel with transport of oxygen via the gas phase.

The reaction rate will then be determined by the diffusion of the faster of the cations, since the transport number of the electronic charge carriers is essentially unity, and, for a sufficiently high porosity, the oxygen gas will always be available everywhere at the reaction front. This reaction mechanism is illustrated in Fig. 6-3 where the differences between this and other reaction mechanisms can clearly be seen. It appears that such a reaction mechanism involving the gas phase probably occurs during the production of ferrites or chromites [37].

Even for compact product layers, gas transport can still occur at a free surface, and this can lead to a unique morphology as shown in Fig. 8-11 [38]. Here again, the electronegative component is transported rapidly via the gas phase, while the electrons are transported through the solid, provided that the solid is an electronic conductor. The reaction product can then grow parallel to the surface. The rate of this growth is controlled by the diffusion of the faster cation, since the slower cation is always available for reaction at the three-phase boundary gas/reaction product/reactant. Therefore, as a rule, the overlap of the reaction product at the surface will always occur in the direction towards the reactant containing the slower cation (B in this case). From the ratio of product layer thicknesses $\Delta x_1/\Delta x_2$ (see Fig. 8-11), the ratio of the diffusion coefficients of the slower and faster cations can be calculated in a manner similar to that described earlier in this section for the case of simple tarnishing reactions. If $D_{O^{2-}} \ll D_{B^{3+}} \ll D_{A^{2+}}$ in the reaction product, then it follows from eqs. (6-22) and (6-23) that:

$$\left(\frac{\Delta x_1}{\Delta x_2}\right) \approx \left(\frac{\bar{k}_1}{\bar{k}_2}\right)^{1/2} = \left(8\frac{\bar{D}_B}{\bar{D}_A}\right)^{1/2} \tag{8-37}$$

The numerical factor 8 in eq. (8-37) reflects the fact that the reaction mechanisms at the surface and in the interior are different, so that the reaction volumes for the passage of one ion equivalent are not the same in each case.

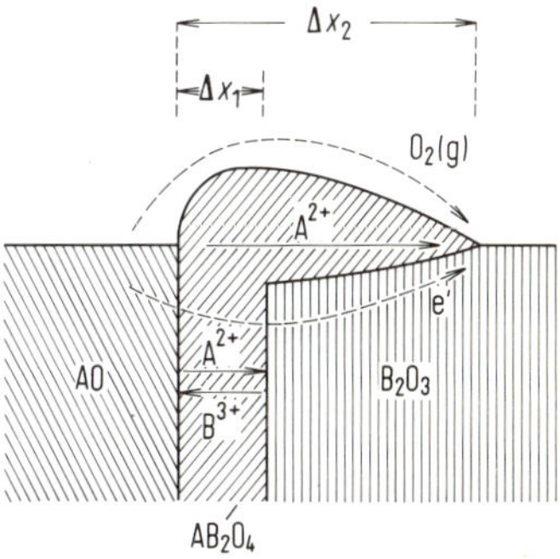

Fig. 8-11. The morphology of the reaction $AO + B_2O_3 = AB_2O_4$ in the vicinity of the surface.

Surface diffusion can also provide a fast transport path, and can thus also give rise to rapid growth along the surface. Although this situation is quite different from that of gas transport which was just described, the resultant morphologies can often be indistinguishable. Particularly striking examples of such phenomena are found for silicate formation reactions as illustrated in Fig. 8-12.

Co_2SiO_4 is a *p*-type semiconductor. Comparison of the measured diffusion coefficient of Co (which is about 10^{-12} cm^2/sec in air at 1300 °C) with the experimentally measured parabolic growth rate constant shows that, for normal growth of the silicate product in region I (see Fig. 8-12), the rate-controlling step is either the diffusion of oxygen ions or the diffusion of silicon ions. In region II, along the edge of the SiO_2, it is the diffusion of Co ions which is rate-determining, since here oxygen is transported through the gas phase, and electron holes move with virtually no resistance through the Co_2SiO_4. Consequently, the product layer in region II is thicker than in region I. Finally, the silicate spreads out sideways across the CoO crystal (region III). This growth occurs as a result of the surface diffusion of a silicon-containing species SiX_{ad} along the surface of the wedge of Co_2SiO_4. The exact nature of this species is not known. If local equilibrium is assumed, then the boundary conditions of the problem are uniquely determined in all three regions. Consequently, the transport equations can be formulated and solved by known methods. The explicit calculation for the wedge-like growth morphology in region III is rather complicated. One of the boundary conditions in this region can be formulated by stating

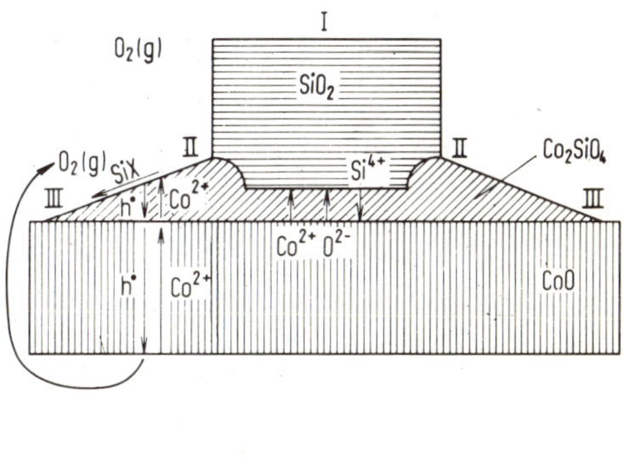

Fig. 8-12. The morphology of the reaction $2\ CoO + SiO_2 = Co_2SiO_4$ in air. SiX is a silicon containing adsorbed particle of unknown composition which diffuses rapidly along the surface.

that the local flux of the rapidly diffusing species SiX_{ad} along the surface of the wedge is reduced locally by the silicate forming reaction with Co^{2+} and O_2(g). Electron holes formed in this reaction diffuse through the wedge toward the CoO/gas surface as shown in Fig. 8-12. Thus, the flux of SiX_{ad} along the surface of the wedge is not locally constant. As a consequence, the silicon activity is lowered towards the end of the wedge, and the reaction rate of the silicate formation is correspondingly decreased.

Interface instability during solid-solid reactions has been observed in the system MgO-SiO_2. Since neither reactants nor the reaction products Mg_2SiO_4 and $MgSiO_3$ are known to have an appreciable range of homogeneity, the phase boundaries ought to be flat in a one-dimensional experiment. Nevertheless, the interface SiO_2/$MgSiO_3$ starts to become uneven right from the beginning of the reaction, and after some time an aggregate two-phase structure is found with a pronounced growth texture.

The explanation is probably due to the fact that 1. SiO_2 as a reactant is not completely homogeneous in structure so that in the very beginning of the reaction a microscopically uneven interface develops, 2. the volume change between reactant and reaction product is appreciable so that a growth form is preferred that minimizes the overall strain, and 3. it is well known that along the surface or interface of silicates a marked increase in diffusivity is found as compared to the bulk diffusivities. The combination of these three effects could lead to a high growth rate of the platelike $MgSiO_3$ reaction product with a marked growth texture as observed.

Many other interesting topochemical reactions could be discussed here. However, the examples which have been given serve to illustrate the fundamental principles.

8.4. Thermal decomposition of solids

Limestone ($CaCO_3$) can be calcined in a kiln to produce solid CaO and CO_2 gas. The reaction is endothermic, and so, in every piece of limestone, heat must be supplied to the reaction site. This heat must be transported through the laminar flowing boundary layer in the gas phase, and through the CaO product layer. Simultaneously, the CO_2 which is produced must be transported away in the opposite direction. The overall process involves a coupled transport of heat and mass. The relationship between the CO_2 partial pressure and the temperature at the reaction site can be determined from thermodynamic data (unless thermodynamic equilibrium is not achieved, in which case additional kinetic data will also be required to determine the relationship). Fuel is burned in the kiln to supply the heat necessary to maintain the reaction.

Another example of a decomposition reaction is the dehydration of magnesium hydroxide according to:

$$Mg(OH)_2 - MgO + H_2O$$

The decomposition of carbonates and hydroxides are among the most thoroughly studied decomposition reactions. The general reaction equation for such processes may be written as:

$$A(s) = B(s) + X(gas) \tag{8-38}$$

In a purely formal sense, this is the inverse of a tarnishing reaction. During a decomposition process, the equilibrium in the reaction eq. (8-38) generally lies far to the right-hand-side. If the reaction product B forms a very porous coating on the outer surface of the reactant A, then the gas which is produced can escape unhindered through the pores into the bulk gas phase. In this case, the phase boundary reaction at the A/B phase boundary will be rate-determining. On the other hand, if a dense layer of B is formed, then the transport of X through this layer will control the overall rate. It can be appreciated that topochemical processes will play an important role here, and so the usual methods of chemical homogeneous and heterogeneous kinetics are not directly applicable in their simplest form. In technological processes, it is desirable that the decomposition rates be high, and so, in such cases, the solution of the coupled heat and mass transport problem is of prime importance. In this book, however, we are mainly concerned with atomistic reaction mechanisms and with the atomistic interpretation of the reaction kinetics. Therefore, as far as possible, we shall deal only with isothermal experiments.

There is no general theory of decomposition reactions. However, an empirical rate curve similar to that shown in Fig. 8-13 is often found.

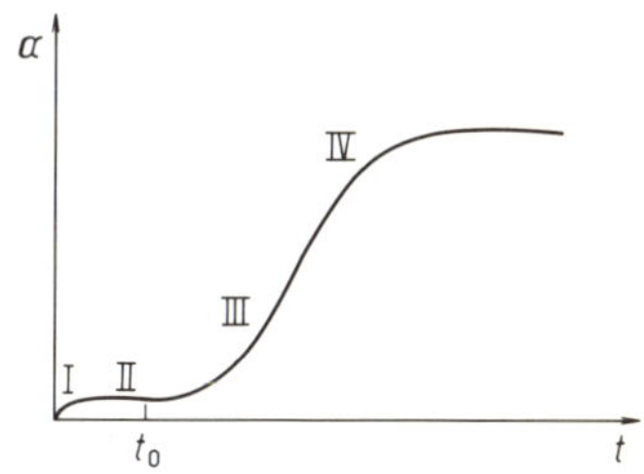

Fig. 8-13. General form of a decomposition curve. The relative yield α is plotted as a function of time.

If the relative yield α is plotted versus time, then four distinct regions can be seen. In region I, a small evolution of gas is observed which typically amounts to about one percent of the total yield. This can be attributed to the desorption of adsorbed gases or to the decomposition of distorted regions of the crystal near the surface. Region II is an incubation period. At time t_0, the rate of decomposition begins to increase rapidly (region III), until an inflection point is reached. Finally, in region IV, the rate continually decreases until the reaction eventually ceases. At this point, the total yield will not necessarily always be 100% ($\alpha = 1$). Decomposition processes have also been observed in which one or more of the regions in Fig. 8-13 do not occur [55].

For almost fifty years, attempts have been made to interpret the experimental results in Fig. 8-13 by an ever increasingly sophisticated phenomenological theory [39]. This theory is concerned with the nucleation and growth of phase B under the assumption that the rate of decomposition is equal to the rate of the phase boundary reaction. This phase boundary reaction rate, in turn, is assumed to be either constant, or to depend uniquely upon the area of the interface already formed [8]. The growth of phase B is generally three-dimensional. Thus, if either the reactant or the reaction product are anisotropic, it is expected that the rate of growth will be different in the direction of each different crystal axis. In certain cases, microscopic observations seem to confirm that the growth rate of B is constant. Whether this supports the postulate that the phase boundary reaction is constant on a submicroscopic scale or not remains an open question.

Up until now, our discussion of solid state reactions has been based upon the assumption of local equilibrium and upon diffusion theory and the associated concepts of defect thermodynamics. From the above introductory remarks, however, we see that, in the case of decomposition reactions, we no longer have this foundation. Furthermore, the decomposition product B is often formed in a highly activated state, such that its free energy is several kcal/mole greater than the standard value. Thus, the thermodynamic foundation also becomes shaky [45]. Therefore, it is not surprising that many decomposition reactions which can be described morphologically and phenomenologically, nevertheless can not be interpreted on a detailed statistical thermodynamic basis.

8.4.1. Kinetics of decomposition of solid phases

From the above remarks we can see that any expression to describe the kinetics of decomposition must be very restricted and qualified. There are two possible approaches. Either we can attempt to generalize a body of experimental data, or else we can start with idealized models and can calculate theoretical decomposition curves which may then be compared to experimental measurements. In the latter case, the parameters of the model should preferably be obtained by independent measurements. Studies involving both approaches have been reviewed [8, 55, 56]. In this section, a few general considerations and expressions will be discussed.

According to the Gibbs phase rule and the reaction equation (8-38), the equilibrium partial pressure of X at the phase boundary A/B is uniquely determined in a quasi-binary system when the total pressure and temperature are fixed. If the reaction proceeds slowly enough so that isothermal conditions prevail, then the deviation from the equilibrium partial pressure p_X (eq) will be the driving force for the reaction, and will thus determine the reaction rate. Decomposition reactions have often been carried out under non-isobaric conditions in order that the kinetics of the reaction might be directly investigated by measurements of the change in pressure in a closed system. However, the experimental reaction kinetics in such cases cannot be easily interpreted, since the experimental parameters are continually changing.

As an example of an isobaric decomposition, let us consider the reaction [46]

$$CaCO_3 - CaO + CO_2(g)$$

If natural calcite crystals are used, then a decomposition curve of the type shown in Fig. 8-13 is observed. The details of the curve will depend upon the deviation of the CO_2 partial pressure from the equilibrium value. Such a decomposition curve with a reproducible incubation period indicates that the rate is controlled by the nucleation and growth of oxide particles.

However, if finely crystalline $CaCO_3$ is used, then the rate of decomposition is found to be proportional to the instantaneous area of the phase boundary $CaCO_3/CaO$. From this it may be concluded that the decomposition kinetics are controlled by a phase boundary reaction which proceeds at a constant rate. In the case of fine-grained material, the nucleation of oxide is no longer a rate-determining kinetic step. Apparently there are always enough active sites for nucleation. The experimental rate constant varies inversely as the partial pressure of CO_2. Since the decomposition process must cease when the CO_2 pressure is equal to its equilibrium value, it follows that the rate equation for the phase boundary reaction must be of the form

$$\mathrm{d}n_B/\mathrm{d}t \propto 1/p_X - 1/p_X(\mathrm{eq})$$

This pressure dependency can be explained as being due to the formation of an activated intermediate product B* (CaO*) at the phase boundary. This intermediate product either reacts back with X to form reactant A again, or else it forms stable crystalline B as end product.

If we now look at the decomposition kinetics of magnesite, we see how little one is able to generalize in such reactions. In the early stages, the yield varies as the square root of the time. Furthermore, the morphology of the MgO which is produced is strongly dependent upon the deviation of the partial pressure of CO_2 from the equilibrium value. The greater is the deviation from equilibrium, the more finely grained is the reaction product. This observation is in agreement with the expected increased rate of nucleation. Of course, the picture may be changed because of sintering and recrystallization processes in the extremely fine-grained reaction product.

If it is assumed that nucleation of phase B is equally probable on all inner and outer surfaces of the crystal of A, and that the rate of the phase boundary reaction for the growth of phase B is constant, then the rate law for the dissociation reaction can be calculated, provided that the nucleation probability is known as a function of time. In the simplest case, there will be N_N possible nucleation sites, each of which has an equal a priori probability of becoming an actual nucleus. If n_N is the number of nuclei already present, then the rate of nucleation is:

$$\mathrm{d}n_N/\mathrm{d}t = k(N_N - n_N) \tag{8-39}$$

or

$$n_N = N_N[1 - \exp(-kt)] \tag{8-40}$$

For small values of kt, only the first term in the power series expansion of the exponential is significant. Eq. (8-40) can then be simplified to:

$$n_N = kN_N t \tag{8-41}$$

This so-called linear nucleation law (k = rate constant for nucleation) presupposes an equal a priori probability of nucleation at all inner and outer surfaces, as well as a slow rate of nucleation. However, if the time required for the thermal decomposition of the crystal is large compared to $1/k$, then n_N will become approximately equal to N_N after only a short period at the beginning of the reaction. That is, nucleation will be complete almost immediately.

If we use eq. (8-41) to describe the nucleation process, and if we take into account the fact that no new nuclei can form in those regions in which A has already decomposed, then, assuming that the rate of the phase boundary reaction is constant during the growth of the nuclei, we obtain the following equation for the relative advancement of the reaction, α, as a function of time:

$$\alpha = 1 - \exp(-at^4) \tag{8-42}$$

where a is a constant which is a function of the number of nucleation sites N_N, the rate constant for nucleation k, and the rate of the phase boundary reaction. The latter may be anisotropic in certain cases. Eq. (8-42) is a special case of the so-called Erofeev equation [40] which has the general form:

$$\alpha = 1 - \exp(-at^n) \tag{8-43}$$

Values of the exponent n other than $n = 4$ will occur if a nucleation law other than that in eq. (8-41) is obeyed, or if the growth of phase B is one- or two-dimensional. It can easily be shown that the essential features of the rate curve following the incubation period, as shown in Fig. 8-13, are well-fitted by the Erofeev equation. Also, it is not difficult to refine the assumptions and the derivation of this equation without altering the physico-chemical concepts upon which it is based. However, to what extent these refined treatments of the phenomenological theory are meaningful will remain an open question as long as the nucleation processes are not independently observed but – as is the case today with few exceptions – are only inferred from the overall decomposition kinetics.

In order to determine the reaction constant a, experimental values of the quantity $f(\alpha) = \left[-\frac{1}{a} \ln (1 - \alpha)\right]^{1/n}$ are plotted versus time, and the value of n which gives the best straight line is sought. In general, correction will have to be made for the incubation time, as shown in Fig. 8-13. There have been several attempts to explain this incubation period. A common hypothesis is that the rate of the phase boundary reaction (i.e. the linear growth rate) for small nuclei is different from that for large particles. The difference is due to distortions which are produced in the immediate neighbourhood of the small nucleus which is growing coherently with the matrix [44]. Once again, this hypothesis cannot be directly confirmed, but can only be inferred from the shape of the decomposition curve $\alpha(t)$. Other explanations are possible. For example, if nucleation initially occurs only along dislocation lines, then a transition from three-dimensional spherical growth to two-dimensional cylindrical growth will occur when the nuclei grow into each other along the dislocation line [42].

Finally, a few remarks should be made regarding decomposition reactions which proceed autocatalytically over a certain time interval. In such cases:

$$\frac{d\alpha}{dt} = b\alpha; \quad \alpha = \alpha_0 \exp bt \tag{8-44}$$

where b is the rate constant. Eq. (8-44) cannot hold for long times, since as $t \to \infty$ it is necessary that $\alpha \to 1$. The equation $d\alpha/dt = b\alpha(1 - \alpha)$ is therefore better than eq. (8-44). Garner [43] has suggested that eq. (8-44) will result from a linear growth of phase B if this phase exists in the form of branches, and if there is a constant probability that the branches will split. Obviously, this model leads immediately to eq. (8-44). More recent hypotheses [41] stress the importance of the strain which the growing product phase B induces in the surrounding matrix of A. This strain can lead to the formation of new dislocations or even to micro-cracks [44]. These dislocations and inner surfaces then act as sites of preferred nucleation, and so an autocatalytic effect results.

A more comprehensive discussion of individual examples of decomposition reactions will not be given here. Experimental measurements can be affected quite markedly by the preparation and handling of the sample as well as by the surrounding gas atmosphere. Consequently, it can be very difficult to achieve reproducibility. Therefore, it is not an easy matter to study the elementary atomic steps of the decomposition reaction. In order to obtain an idea of the difficulties involved in interpreting the experimental results, reference may be made to review articles [8, 55, 56], and especially to the work of Haul [47] who has made a very thorough study of the decomposition of dolomite.

8.5. Literature

General Literature:

[1] P. Kofstad, High-Temperature Oxidation of Metals, John Wiley and Sons, Inc., New York 1966.
[2] K. Hauffe, Oxidation of Metals, Plenum Press, New York 1965.
[3] O. Kubaschewski and B. E. Hopkins, Oxidation of Metals and Alloys, 2nd ed., Butterworths, London 1962.
[4] Th. Wolkenstein, Elektronentheorie der Katalyse an Halbleitern, Deutscher Verlag der Wissenschaften, Berlin 1964.
[5] E. Spenke, Elektronische Halbleiter, 2nd ed., Springer-Verlag, Berlin 1965.
[6] H. Pfeiffer and H. Thomas, Zunderfeste Legierungen, 2nd ed., Springer-Verlag, Berlin 1963.
[7] H. Kaesche, Die Korrosion der Metalle, Springer-Verlag, Berlin 1966.
[8] D. A. Young, Decomposition of Solids, Pergamon Press, Oxford 1966.

Special Literature:

[9] J. Bénard et al., Z. Elektrochem. *63*, 799 (1959).
[10] J. B. Holt and L. Himmel, J. Electrochem. Soc. *116*, 1569 (1969).
[11] N. B. Pilling and R. E. Bedworth, J. Inst. Metals *29*, 529 (1923).
[12] C. Wagner, Z. phys. Chem. *B21*, 25 (1933).
[13] D. L. Douglass and C. Wagner, J. Electrochem. Soc. *113*, 671 (1966).
[14] C. Wagner, Acta Met. *17*, 99 (1969).
[15] E. Fromm, Z. Metallkunde *57*, 60 (1966).
[16] C. Wagner and K. Grünewald, Z. phys. Chem. *B 40*, 455 (1938).
[17] W. Jost, Diffusion und chemische Reaktion in festen Stoffen, Steinkopff-Verlag, Dresden 1937, p. 149.
[18] K. Hauffe and H. Pfeiffer, Z. Metallkunde *44*, 27 (1953).
[19] F. S. Pettit and J. B. Wagner, Acta Met. *12*, 35 (1964).
[20] N. F. Mott, J. chim. phys. *44*, 172 (1947); Progr. Phys. *12*, 163 (1949).
[21] K. Hauffe, Oxidation of Metals, Plenum Press, New York 1965, p. 126ff.
[22] G. K. Boreskov, Discuss. Faraday Soc. *41*, 263 (1966).
[23] C. Wagner, Corrosion Science *10*, 641 (1970).
[24] C. Wagner, Z. Elektrochem., Ber. Bunsenges. phys. Chem. *63*, 772 (1959).
[25] C. Wagner, J. Electrochem. Soc. *99*, 369 (1952).
[26] O. Kubaschewski and O. von Goldbeck, J. Inst. Metals *76*, 255 (1949).
[27] R. A. Rapp, Acta Met. *9*, 730 (1961).
[28] J. L. Meijering and M. J. Druyvesteyn, Philips Res. Repts. *2*, 81, 260 (1947).
[29] C. Wagner, Corrosion Science *9*, 91 (1969).
[30] N. Birks and H. Rickert, J. Inst. Metals *91*, 308 (1962/63).
[31] B. D. Lichter and C. Wagner, J. Electrochem. Soc. *107*, 168 (1960).
[32] P. Delahay, M. Pourbaix and P. van Rysselberghe, J. Electrochem. Soc. *98*, 57, 65, 101 (1951).
[33] G. Kortüm, Lehrbuch der Elektrochemie, 3rd ed., Verlag Chemie, Weinheim 1962, p. 531.
[34] K. J. Vetter, Z. phys. Chem. NF *4*, 165 (1955).
[35] C. Ilschner-Gensch and C. Wagner, J. Electrochem. Soc. *105*, 198 (1958).
[36] J. H. Eriksen and K. Hauffe, Z. phys. Chem. NF *59*, 326 (1968).
[37] P. Reijnen in J. W. Mitchell, R. C. DeVries, R. W. Roberts and P. Cannon, Reactivity of Solids, Proc. 6. Intern. Symp. 1968, Wiley-Interscience, New York 1969, p. 99.
[38] H. Rickert and C. Wagner, Ber. Bunsenges. phys. Chem. *66*, 502 (1962).
[39] J. Y. MacDonald and C. N. Hinshelwood, J. chem. Soc. *127*, 2764 (1925).
[40] B. V. Erofeev, Compt. Rend. Acad. Sci. URSS *52*, 511 (1946).
[41] P. W. M. Jacobs in T. J. Gray and V. D. Frechette, Materials Science Research, Vol. 4, Kinetics of Reactions in Ionic Systems, Plenum Press, New York 1969, p. 37.
[42] D. A. Dominey, H. Morley and D. A. Young, Trans. Faraday Soc. *61*, 1246 (1965).
[43] W. E. Garner and H. R. Hailes, Proc. Roy. Soc. Ser. *A 139*, 576 (1933).
[44] E. G. Prout and F. C. Tompkins, Trans. Faraday Soc. *40*, 488 (1944).

[45] R. Fricke and J. Lüke, Z. Elektrochem. *41*, 174 (1935).
[46] E. Cremer and W. Nitsch, Z. Elektrochem., Ber. Bunsenges. phys. Chem. *66*, 697 (1962).
[47] R. Haul et al., Z. anorg. allg. Chem. *269*, 120 (1952); *281*, 199 (1955).
[48] G. Tammann, Z. anorg. allg. Chem. *111*, 78 (1920).
[49] T. P. Hoar and L. E. Price, Trans. Faraday Soc. *34*, 867 (1938).
[50] H. J. Grabke, Ber. Bunsenges. phys. Chem. *69*, 48 (1965).
[51] G. C. Wood and I. G. Wright, Corrosion Science *5*, 841 (1965).
[52] Freiberger Forschungshefte B 142, Dispersionshärtung, VEB Deutscher Verlag für Grundstoffindustrie, Leipzig 1969.
[53] C. Wagner, Werkstoffe und Korrosion *21*, 886 (1970).
[54] W. Schwenk, Stahl und Eisen *10*, 535 (1969).
[55] P. W. M. Jacobs and F. C. Tompkins in W. E. Garner, Chemistry of the Solid State, Academic Press, London 1955, p. 184.
[56] W. E. Garner, Chemistry of the Solid State, Academic Press, London 1955.
[57] A. D. Pelton, H. Schmalzried and C. D. Greskovich, Ber. Bunsenges. phys. Chem. *76*, 543 (1972).
[58] H. Schmalzried in G. M. Schwab, Reactivity of Solids, Proc. 5. Intern. Symp. 1964, Elsevier Publ. Comp., Amsterdam 1965.
[59] W. W. Smeltzer, Battelle Colloquium on Defects and Transport in Oxides, Columbus 1973.

9. Some technologically important solid state reactions

In this chapter we shall discuss some solid state reactions which are of technological importance. We shall not be too concerned with providing a quantitative treatment. Technological processes are often much too complex to permit us to give a complete quantitative description of the overall physico-chemical course of the reaction. Instead, we shall apply the previously discussed concepts in two ways. Firstly, we shall attempt to separate out the elementary reaction steps and to describe these steps in terms of fundamental atomic parameters. Secondly, we shall try to describe the processes on the basis of simple limiting cases which can be tested in practice. The examples to be discussed have been chosen somewhat arbitrarily. However, many different areas of solid state chemistry are covered by these examples, and the wide range of applicability of the basic concepts of solid state reactions is well illustrated.

9.1. Formation of the microstructure in ferrites and titanates

In the simplest case, oxide-ferrites are double oxides of the form $MeFe_2O_4$. Examples are $MgFe_2O_4$, $NiFe_2O_4$, or $Zn\,Fe_2O_4$. These compounds crystallize in the spinel structure. $ZnFe_2O_4$ is a normal spinel, $NiFe_2O_4$ is an inverse spinel, and $MgFe_2O_4$ is a mixed spinel in which the temperature dependent distribution of cations upon the tetrahedral and octahedral sites lies between the two limiting cases of a normal and an inverse spinel. Electron conduction or electron hole conduction can occur in these materials, and this conduction is influenced by chemical composition, oxygen partial pressure, and temperature. Furthermore, these materials exhibit macromagnetic properties as a result of interactions between the elementary magnetic dipoles of the cations. Because of these electrical and magnetic properties, the group of substances has become indispensible to the electronics industry today. It can be truthfully said that the entire technology of high frequencies between 10^3 and 10^{11} Hz would be entirely different were ferrites not available for use as control or circuit elements.

Titanates are double oxides of the form $MeTiO_3$ or Me_2TiO_4. Barium titanate $BaTiO_3$ and its solid solution crystals with other titanates are especially well-known. $BaTiO_3$ crystallizes in the perovskite structure. Its technical importance results from its ferroelectric and associated piezoelectric properties, its high dielectric constant at room temperature, and the interesting semiconducting properties which it exhibits when doped [13]. The remarkable temperature dependence of the electrical resistance of such doped material (the temperature coefficient can be metal-like) is used to advantage in control and circuit devices.

These materials thus possess a combination of technically interesting electrical and magnetic properties which are dependent upon the chemical composition, the crystallographic structure, and the independent thermodynamic variables such as temperature and oxygen partial pressure, as well as upon the microstructure (i. e. upon grain size, dislocation distribution, porosity, and inclusions). In order to obtain a material with a certain technologically desirable property or a combination of different properties, one must be concerned not only with cation distribution and defect concentrations, but also with obtaining an optimal microstructure with the proper grain size distribution and porosity. Therefore, it is necessary not only to understand the physical properties and the defect thermodynamics of the system, but also to be able to control the formation of the microstructure. This end can be achieved by control of the formation reaction proper ($MeO + Fe_2O_3 = MeFe_2O_4$, $BaO + TiO_2 = BaTiO_3$, or $BaCO_3 + TiO_2 = BaTiO_3 + CO_2$). The reaction, which usually occurs as a powder reaction, can be influenced by temperature, oxygen partial pressure, and doping (see section 6.3.2). As well, however, the grain size, pore size, and porosity can be regulated by means of controlled sintering and grain growth. We shall now present a brief description of the optimal microstructure for certain properties in this group of substances. Following this, we shall discuss the fundamentals of sintering and grain growth, and their relationship to the basic concepts of solid state reactions.

9.1.1. Optimal microstructure

First of all, let us discuss titanates. If we are interested primarily in the piezoelectric properties of the titanate, then we are concerned with the ability of the tetragonal domains to line up with one another. These domains, which occur below the Curie point, are about 1 μ in width. Since the crystallites are hemmed in by their neighbours, they can only partially submit to the polarization-induced shape change (electrostriction). However, if the crystallites are large enough ($> 10\ \mu$) so that different domain systems can occur within one crystallite, then a partial compensation for the changes in domain shape is possible. This is not possible for crystallites $< 1\ \mu$ which consist of only one domain. Optimal piezoelectric properties are thus expected in dense specimens with grain sizes between 10 and 20 μ.

Titanates can also be used as dielectrics in condensers to give relative dielectric constants as high as 4000. For this application, we seek to suppress piezoelectric effects and ferroelectric hysteresis losses. This can best be achieved, according to the ideas above, by producing high density, very fine-grained material with grain sizes $< 1\ \mu$.

As mentioned previously, $BaTiO_3$, when doped with ions of higher valence, exhibits interesting semiconducting properties [14]. A large increase in electrical resistance occurs above the Curie point. It has been demonstrated that the semiconducting properties of this titanate are connected with grain boundary resistivity effects and with the presence at the grain boundaries of oxygen from the gas phase. Consequently, the optimal microstructure is obtained in a polycrystalline and porous sample in which the diameter of the crystallites is small.

Similar considerations apply to the production of ferrite samples with prescribed electrical and magnetic properties. The motion of the boundaries of the elementary magnetic domains (the so-called Bloch walls) during magnetization is blocked by pores or grain boundaries. In very fine-grained samples, the coercive force is markedly increased over that for coarse-grained or monocrystalline material. Once again, we see how the physical properties of the substance depend upon the microstructure [1]. In the following section, we shall discuss the possibility of influencing the microstructure by means of transport processes in the solid state.

9.1.2. Formation of the microstructure during sintering

The aforementioned electrical and magnetic control and circuit elements must be produced in certain geometrical shapes either by means of mechanical finishing of the unfinished piece, or by shaping and sintering during the production process. The shaping is accomplished by dry pressing, by slip-casting, or by shaping ceramic paste with the help of an organic binder [15].

The unsintered sample should be as dense as possible. A homogeneous pressure distribution is most important in this regard. If the sample is then brought to a high enough temperature, sintering will begin. The strength and solidity of the sample will increase, the porosity will decrease, and grain growth will often occur. The temperature must be chosen so that a uniform contraction and compaction takes place by sintering, but so that no other deformation occurs.

If no molten phase appears during the sintering process, then the compaction (decrease in number and size of pores) occurs through mass transport in the solid phase, and the grain growth takes place by a rearrangement of lattice atoms at the moving grain boundaries. Well-known examples of this type of sintering from the field of powder metallurgy are the production of compact tungsten, molybdenum, and sintered iron. Oxide ceramic products are also made in this way.

In order to analyze the sintering process, we must first inquire what the driving force for the mass flux is, and also which kind of transport processes (and consequently which transport coefficients) determine the kinetics of sintering. The free energy of a given amount of a single-phase substance will be a minimum when the substance exists as a perfect single crystal. Therefore, the driving force for the sintering process is the decrease in surface free energy which occurs as the surface area of the polycrystalline aggregate is reduced. Simultaneously, the crystallites will grow in order to decrease the interfacial energy of the grain boundaries. It should be noted that, in solids, the motion of the atoms is relatively restricted compared to liquid phases. Hence, the surface or interface tension and the specific free surface or interface energy cannot be directly equated [16].

In porous media, several modes of transport are possible [17]. As long as the pores are connected with each other and change only in shape during the first stages of sintering, then gas diffusion or Knudsen flow in the pores, as well as surface diffusion, are possible. In advanced stages of sintering, when the pores are closed off, the process continues via crystal volume diffusion, grain boundary diffusion, and diffusion along dislocations. In these latter stages, the pores become filled up. This can only occur as a result of a mass flux to the pores from a source on an inner or outer surface. If only volume diffusion is considered, then this means that interstitial atoms diffuse to the pores or that vacancies diffuse away from the pores, either from or to the outer surface, or from or to low- or high-angle grain boundaries and dislocations.

Generally speaking, in ionic crystals the defects possess an effective electric charge relative to their surroundings. Consequently, the defect fluxes are coupled. Defects with like charges diffuse in opposite directions, or else defects with opposite charges diffuse in the same direction. The latter case is the most common in the case of sintering of ionic crystals. Since grain boundaries serve as sources and sinks for the particle fluxes, it can immediately be appreciated that fine-grained material will have better sintering properties than coarse-grained material. At the same time as the material is becoming more compact because of the filling-up of pores, grain growth is occurring, since this process also decreases the interfacial free energy (in this case, the grain boundary energy).

Consequently, both the geometry and the density of sources and sinks for the sintering reaction are continually changing. When we view the entire sintering process in all its complexity, we can readily understand that a complete quantitative treatment is virtually impossible [17]. Therefore, in the following, we shall simply discuss the concepts and sketch out the equations necessary for a quantitative analysis.

In order to simplify the discussion, let us consider a model in which cylindrical particles of equal radius are piled up in a close-packed array. Furthermore, we distinguish three stages of sintering in the sense of limiting cases. In the first stage, the porosity remains constant. However, the pores assume that shape which has the minimum surface area. That is, they take on a circular cross-section as shown in Fig. 9–1. Rearrangement of the particles can occur by any of the previously mentioned transport paths. Plastic flow can also play a role here. Depending upon which transport path is assumed to be predominant, quite different rate laws for the growth of the contact surface between the cylinders will result (see Fig. 9-1).

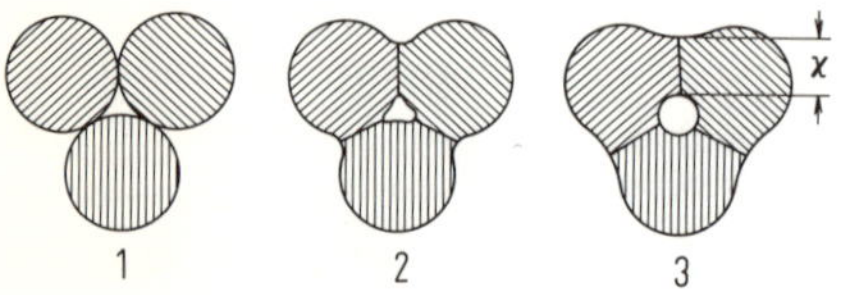

Fig. 9-1. The first stage of sintering. The pores assume the shape of minimum surface area.

If x is the width of the contact area, then the rate law over a certain time interval will be of the form $x = k t^n$, where $n < 1$.

The second stage may be defined as follows: Grain boundaries emanate from every pore to the neighbouring pores. Grain growth has not yet occurred (see Fig. 9-2).

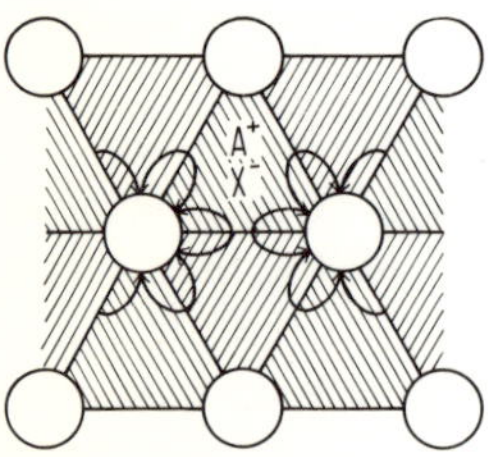

Fig. 9-2. The second stage of sintering: Matter is transported from grain boundaries and surfaces to the pores. A^+ = cation, X^- = anion.

Because of the negative curvature of the surface of a small pore, the chemical potential of the substance is less there than it is at planar interfaces such as grain boundaries or outer surfaces. The resultant potential gradient gives rise to a mass flux towards the pore. The flux will be proportional to the appropriate transport coefficient. Both grain boundary diffusion and volume diffusion can occur. In the following, we shall assume that volume diffusion is predominant, but the calculations can just as easily be performed for the case of grain boundary diffusion. In practice, both mechanisms are active. In the first approximation, a pore can be considered to be small relative to the distance between the pore and the flux source, so that the chemical potential difference $\Delta\mu$ changes with the pore radius r_P according to the Gibbs-Thomson equation:

$$\Delta\mu = \frac{V_m \gamma_0}{r_P} \tag{9-1}$$

where γ_0 is the interfacial free energy of the substance. For cylindrical pores, in contrast to spherical pores, there is no factor 2 in the numerator of eq. (9-1). It should be noted that, in this expression, no distinction is made between the interface crystal- (interior of pore), and a grain boundary. For the sintering of a metal, there is no conceptual difficulty involved in the application of eq. (9-1). For the sintering of a compound AX, however, the condition of electroneutrality requires that there be a coupled flux of ions A^+ and X^-. This must be taken into account in the calculations. By eliminating the diffusion potential from the general transport equations (5-13), we obtain the equation:

$$j = j_{A^+} = j_{X^-} = -\frac{D_A D_X}{D_A + D_X} \cdot \frac{1}{RTV_{AX}} \cdot \text{grad}\, \mu_{AX} \qquad (9\text{-}2)$$

where V_{AX} is the molar volume of AX. For the practical evaluation of eq. (9-2), etc., the component diffusion coefficients D_i may be replaced by tracer diffusion coefficients D_i^* if correlation effects are neglected. Because of the cylindrical symmetry, the gradient will be relatively steep in the neighbourhood of the pore. Thus, to a good approximation we can write:

$$\left(\frac{\partial \mu_{AX}}{\partial r}\right)_{r=r_P} = \alpha \frac{\Delta \mu_{AX}}{r_P} \qquad (9\text{-}3)$$

where α is a numerical factor of the order of unity. The total flux per unit length of the pore is then given by eqs. (9-1) and (9-3) as:

$$|I| = 2\pi r_P j \approx \frac{2\pi\alpha\gamma_0(AX)}{RT} \cdot \frac{D_A D_X}{D_A + D_X} \cdot \frac{1}{r_P} \qquad (9\text{-}4)$$

This flux is responsible for the decrease in the volume of the pore. If V_P is the volume of the pore per unit length, then:

$$-\frac{dV_P}{dt} = |I|\, V_{AX} \approx \frac{2\pi\alpha\gamma_0(AX)}{RT} \cdot \frac{D_A D_X}{D_A + D_X} \cdot \frac{V_{AX}}{r_P} \qquad (9\text{-}5)$$

By integrating this equation, we find that, under the stringently limiting assumptions which we have made, the volume of the pore decreases proportionally to $(1 - \beta t)^{2/3}$. The factor β contains the molar volume V_{AX}, the effective diffusion coefficient $D_A D_X/(D_A + D_X)$, and the surface free energy $\gamma_0(AX)$. Therefore, the factor β can, in principle, be calculated in terms of known quantities. Once again, we note that the more slowly diffusing ion is rate-determining.

In these calculations concerning the second stage of sintering, we have assumed that the pores are intersected by grain boundaries. However, not all pores are of this type. In Fig. 9-3 is shown schematically the distribution of pores which is found in the third stage of sintering after grain growth has occurred as a result of recrystallization. In this stage, the pores are found within the crystallites.

It is evident once again from Fig. 9-3 that the grain boundaries act as vacancy sinks (or as sources for atoms or ions), since the pores have completely disappeared in the immediate vicinity of the grain boundaries. The quantitative treatment is analogous to that just given above. Finally, in this third stage of sintering, it will be observed that the large pores within the crystallites will grow at the expense of the smaller pores because of the difference in chemical potentials in the neighbourhoods of the small and large pores. This leads to a sort of "Ostwald ripening" of the pores (see section 7.3.4), but no overall increase in density of the porous material results. A density increase only occurs when there is a mass flux from sites of repeatable growth such as grain boundaries or outer surfaces. The growth of the larger pores at the expense of the smaller ones is called parasitic pore growth. This process can be handled quantitatively by the methods of section 7.3.4.

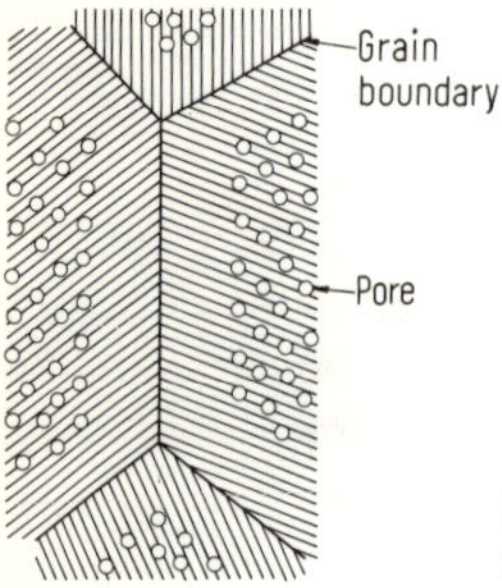

Fig. 9-3. The distribution of pores in the vicinity of the grain boundaries during the third stage of sintering.

In many cases, the addition of a small amount of heterovalent ions can drastically alter the rate of sintering of a pressed specimen of an ionic material. As long as the volume diffusion of one type of ion of the matrix is rate-determining, then this effect can be ascribed to the alteration of the defect equilibria which is brought about by the heterovalent addition. The change in defect concentrations in the cationic and anionic sublattices will result in a change in the chemical diffusion coefficient $\tilde{D} = D_A D_X/(D_A + D_X)$ in eq. (9-4). Now, the concentrations of the defects which are responsible for the ion transport are coupled through defect equilibria, such that the concentration of one type of ionic defect increases when that of the other type decreases, and vice versa. Therefore, it can be easily seen that, in general, $\tilde{D}$ will go through a maximum at some particular concentration of heterovalent addition. A similar situation occurs in the case of ternary compounds, for example in case of sintering of spinels. For instance, the rate of sintering of an aluminate $Me^{II}_{1-x}Al_{2+2/3 \cdot x}O_4$ will depend upon the composition variable x. A change in this composition variable x is equivalent to a change in the amount of heterovalent additions to AX as discussed above. A good example of the strong dependence of the rate of compaction during sintering upon the deviation from the stoichiometric composition of a ternary spinel is shown in Fig. 9-4 [18].

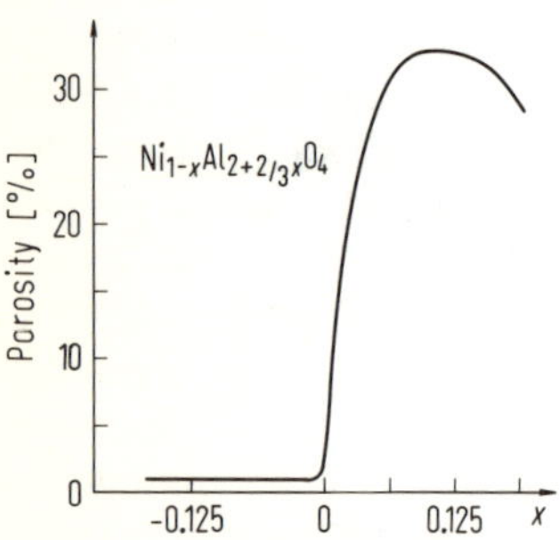

Fig. 9-4. Porosity of a sintered spinel $Ni_{1-x}Al_{2+\frac{2}{3}x}O_4$ as a function of x after the same sintering treatment: 24 hours at 1600 °C in oxygen.

If the sintering process is not controlled by volume diffusion, then these conclusions are no longer valid. Grain boundary diffusion and segregration of the heterovalent addition at the grain boundaries as well as secondary grain growth during sintering can completely change the situation. The details of these processes are described in the special literature [2, 3, 17]. It is no exaggeration to say that sintering is one of the most intensely discussed processes in the area of porous metals and ceramics today.

9.2. Solid state galvanic cells

9.2.1. General introduction

In the preceding chapters, we have frequently required values of the free energies of solid state reactions ΔG, and of the component activities a_i in solid phases. In many cases, these quantities can readily be measured at high temperatures by the use of solid state galvanic cells. Or, if the value of ΔG is known, then galvanic cells can be used to determine ionic transport numbers. Furthermore, solid state cells can be used as fuel cells to produce electrical energy. Conversely, if electrical energy is supplied, then solid state cells can be used to decompose compounds electrochemically into their constituents. For example, the breathing air in space ships could be regenerated in this way. Finally, solid state galvanic cells can be used for kinetic measurements involving solid phases. Examples of all these different applications will be given in this section. It will be shown that all the concepts and formulae of defect thermodynamics and transport theory which have been discussed in previous chapters are applicable to the quantitative treatment of solid state galvanic cells.

In chapters 6 and 7 we discussed the processes which occur when two solid reactants are placed in contact with each other and allowed to react so as to lower the free energy of the system. For example, a product layer of Fe_3O_4 is formed between "FeO" and Fe_2O_3. The quotation marks about the wüstite FeO are to remind us that the stoichiometric compound does not actually exist. Wüstite always possesses a metal deficit [23] (see section 9.4.1). In the following discussion, the quotation marks will be omitted. The reaction between FeO and Fe_2O_3 can also be carried out with an ionic filter inserted between the two reactants, where the conduction in the filter is purely ionic, and the transport number of one of the components of the reactants is unity. Such a solid ionic filter, which is suitable for the above reaction, is ZrO_2 which has been doped with an alkaline earth metal oxide. This material conducts solely by the transport of oxygen ions [24]. The experimental arrangement is shown in Fig. 9-5.

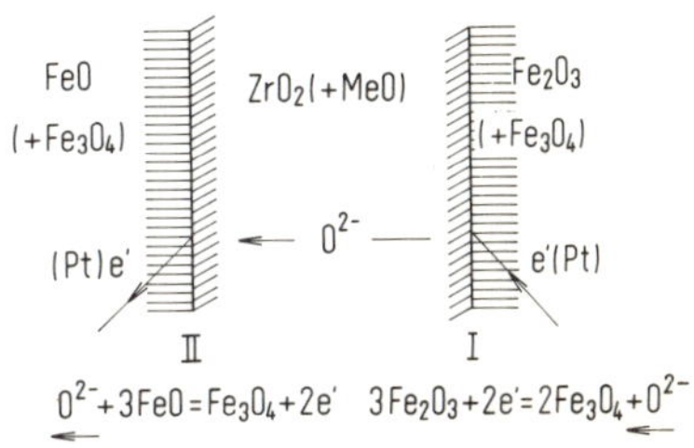

Fig. 9-5. The experimental arrangement and the phase boundary reactions of the solid state galvanic cell: FeO, Fe_3O_4/ZrO_2 (+ MeO)/Fe_2O_3, Fe_3O_4 with ZrO_2 (+ MeO) as solid electrolyte.

In each electrode compartment there is a two-phase mixture: FeO/Fe_3O_4 and Fe_2O_3/Fe_3O_4. Consequently, by the Gibbs phase rule, the reactants FeO and Fe_2O_3 will be invariant. In order for the overall reaction $FeO + Fe_2O_3 = Fe_3O_4$ to take place (i.e. in order for wüstite to be oxidized to magnetite and for hematite to be reduced to magnetite), it is necessary that oxygen ions travel from right to left through the oxygen ion filter $ZrO_2(+MeO)$. This ionic filter may be called the electrolyte. As illustrated in Fig. 9-5, the phase boundary reactions which occur at phase boundaries I and II are:

$$\begin{aligned} &3\,Fe_2O_3 + 2\,e' = 2\,Fe_3O_4 + O^{2-} \qquad &(I)\\ &O^{2-} + 3\,FeO = Fe_3O_4 + 2\,e' \qquad &(II)\end{aligned} \tag{9-6}$$

so that the overall reaction which occurs on the passage of 2 Faradays is given by the sum of the two partial reactions at the phase boundaries as :

$$3\,FeO + 3\,Fe_2O_3 = 3\,Fe_3O_4, \quad \Delta G^0 \tag{9-7}$$

It can be seen that the solid state reaction formulated in eq. (9-6) is electrochemical in nature. In order for the reaction to take place, electrons e′ must be supplied to phase boundary (I), and an equal number of electrons must be removed from phase boundary (II) (cathodic and anodic partial processes). This can be achieved most simply by short-circuiting the cell with a metallic connection between the two-phase boundaries. If this metallic connection is broken, then the reactions (9-6) will proceed only until electrochemical equilibrium is reached. That is, the reactions proceed by consuming electrons at the anode I and by producing electrons at the cathode II until an electric field is built up which prevents the further diffusion of oxygen ions in the electrolyte. Let E be the voltage which is thus set up between the two electrodes. Suppose, now, that with the system in this equilibrium state, and with all intensive parameters held constant, a vanishingly small amount of material δ (expressed in moles) is permitted to pass through the electrolyte. The change in free energy is $(\partial G/\partial \xi)\delta = \Delta G \delta$, where ξ is the measure of the relative advancement of the overall reaction. For one formula amount of reaction, ξ varies from 0 to 1, and at the same time, zF Faradays of electrical charge are displaced. Therefore, when an amount δ of material is transported, the charge which is displaced is δzF, and the corresponding electrical work is δzFE. At electrochemical equilibrium, the two work terms above are equal, and so:

$$\Delta G = -zFE, \quad E = -\frac{\Delta G}{zF} \tag{9-8}$$

For reaction (9-7), z is equal to 2, since each partial electrode reaction in eq. (9-6) involves two electrons. In general, z is the number of equivalents of electrical charge which are transported when one formula amount (in moles) of overall cell reaction takes place.

We can also think of ΔG as the reversible work required to displace $z/4$ moles of O_2 out of electrode compartment I with partial pressure $p^{I}_{O_2}$ into electrode compartment II with partial pressure $p^{II}_{O_2}$. That is, $\Delta G = \frac{RT}{2} \ln (p^{II}_{O_2}/p^{I}_{O_2})$, or:

$$E = \frac{RT}{2\,zF} \ln \frac{p^{I}_{O_2}}{p^{II}_{O_2}} = \frac{\mu^{I}_{O_2} - \mu^{II}_{O_2}}{2\,zF} \tag{9-9}$$

Thus, the open-circuit voltage E (electromotive force, emf) of the above galvanic cell is a measure of the ratio of the oxygen partial pressures at each side of the electrolyte. For the experiment shown in Fig. 9-5, this measured ratio is the ratio of the oxygen partial pressures over the nonvariant two-phase mixtures Fe_2O_3/Fe_3O_4 (I) and FeO/Fe_3O_4 (II).

The general conclusion to be drawn from this specific examples is that solid state galvanic cells with solid electrolytes can be used primarily to measure free energies of reactions. From this, it is often possible to deduce the difference in chemical potentials (or the ratios of activities) of components of the participating phases.

The situation can also be formulated as follows. The combination of two coexisting solid phases of a binary system with an inert metallic electronic conductor (for example, Pt/Me, MeO) will possess a certain definite oxygen partial pressure at a given temperature because of the equilibrium condition for the reaction $Me + 1/2\,O_2 = MeO$. Therefore, this combination can be thought of as a reference electrode. If this electrode is used in conjunction with an auxiliary oxygen-ion conducting electrolyte, then we have a probe which can be used to measure unknown oxygen potentials. Such probes are widely used technologically. For example, they may be used to follow the course of the deoxidation of liquid steel as deoxidant is added [25]. We can thus immediately see a second important application of solid state galvanic cells: If electrochemical equilibrium is attained rapidly enough at the measuring electrode, then the kinetics of time-variant processes can be followed by the measurement of component activities at the electrode.

Our final example of the application of galvanic cells with solid electrolytes is in the field of fuel cells. A fuel cell is defined as an electrochemical device in which the free energy change during the oxidation of a conventional fuel (carbon, methane, CO, H_2) is continuously converted directly into electrical energy [6]. Since a fuel cell must provide electrical energy to an external load, kinetic problems arise at the phase boundaries. Ohmic losses also occur, mainly in the electrolyte. Consequently, the useful voltage of a fuel cell is smaller than the electromotive force E. The first steps towards the development of a workable fuel cell with solid electrolytes were taken long ago by Nernst and Haber [26]. Systematic studies were carried out by Baur and coworkers [27]. From their own experience, these authors were skeptical of the possibility of developing fuel cells with liquid electrolytes. Following the fundamental studies of ZrO_2-based oxygen-ion conducting electrolytes, an intensive effort began after the second world war in many parts of the world to devise useful fuel cells based upon these electrolytes [7]. The great breakthrough has yet to be made, however, since the technological requirements of high current density and invariance of the electrodes and of the electrolyte during use at the necessary high temperatures have up to now only been partially fulfilled.

Before discussing further interesting applications of solid state galvanic cells, let us first examine the thermodynamic and electrical processes which occur in these cells in the light of the general transport theory of solids. We shall then be better able to understand the applications.

9.2.2. Application of the general transport theory to solid state galvanic cells

Let us once again consider, as a specific example, the cell

$$Pt/FeO,\ Fe_3O_4/ZrO_2(+MeO)/Fe_3O_4,\ Fe_2O_3/Pt \tag{I}$$

This cell, which was discussed previously, is shown in Fig. 9-5. We shall examine how the chemical, electrical (Galvani-), and electrochemical potentials (μ_i, ϕ, η_i) vary across this cell.

The knowledge which we gain can then be directly applied to other cells. According to the Gibbs phase rule, the chemical potentials of all particles in the two-phase electrode compartments are constant. The two-phase mixtures are electronic conductors, and so the transport number of the electronic charge carriers is unity. Therefore, the electrical and electrochemical potentials of all partners in the electrode compartments are constant.

According to eq. (5-13), the following transport equations apply within the solid electrolyte:

$$j_{O^{2-}} = -\frac{c_{O^{2-}} D_{O^{2-}}}{RT} \cdot \frac{d}{dx}(\mu_{O^{2-}} - 2\,F\phi) \qquad (9\text{-}10)$$

$$j_{e'} = -\frac{c_{e'} D_{e'}}{RT} \cdot \frac{d}{dx}(\mu_{e'} - F\phi) \qquad (9\text{-}11)$$

The transport number $t_{O^{2-}}$ of the oxygen ions in the electrolyte is assumed to be unity. The transport numbers of the cations are negligible for this material. Furthermore, at open circuit the total current density $I = \sum z_i F j_i$ in the electrolyte must vanish. From these two conditions it follows that $c_{O^{2-}} D_{O^{2-}} \gg c_{e'} D_{e'}$, and therefore, that:

$$\frac{d}{dx}(\eta_{O^{2-}}) = \frac{d}{dx}(\mu_{O^{2-}} - 2\,F\phi) \simeq 0 \qquad (9\text{-}12)$$

The chemical potential of the oxygen ions in the electrolyte is constant because of the large degree of ionic disorder (10% anion vacancies when 10% MeO is added as dopant). Therefore, from eq. (9-12), it may be seen that the electrical potential ϕ is constant in the electrolyte. The situation is illustrated in Fig. 9-6.

The oxygen potentials $\mu_{O_2}^{II}$ and $\mu_{O_2}^{I}$ at either phase boundary of the electrolyte are different, and are fixed by the two-phase mixtures FeO/Fe_3O_4 and Fe_3O_4/Fe_2O_3 through the equilibrium conditions of the reactions:

$$2\,Fe_3O_4 + 1/2\,O_2 = 3\,Fe_2O_3$$

and

$$3\,FeO + 1/2\,O_2 = Fe_3O_4$$

For the reaction:

$$1/2\,O_2 + 2\,e' = O^{2-}$$

the following equilibrium condition may be written:

$$1/2\,\mu_{O_2} + 2\,\mu_{e'} = \mu_{O^{2-}}$$

However, as discussed above, $\mu_{O^{2-}}$ is constant in the electrolyte. Therefore, the chemical potentials (and concentrations) of the electrons at either phase boundary will be different. Since the electrodes are at least in part reversible to electrons, a small diffusive flux of electrons is operative through the electrolyte of the galvanic cell. The variation of the potentials and concentrations across the galvanic cell is schematically illustrated in Fig. 9-6.

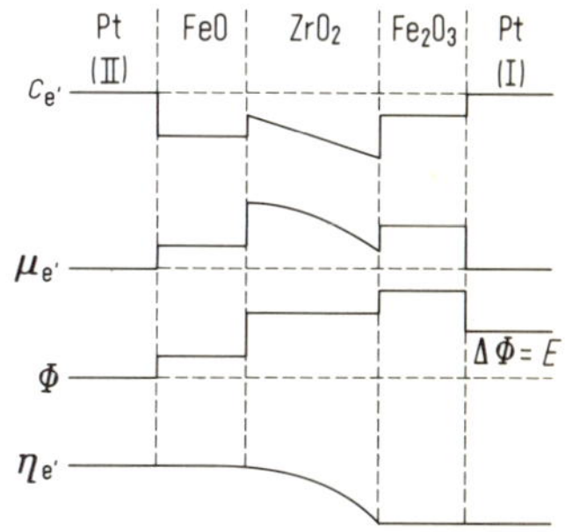

Fig. 9-6. Variation of the electrical, chemical, and electrochemical potentials in the solid state galvanic cell illustrated in Fig. 9-5. The variation of the concentration of electrons is also shown. The transport number of the ions, t_{ion}, in the electrolyte is equal to one.

The jump in electrical potential at each individual phase boundary is unknown. However, the difference in Galvani potential $E = \phi_I - \phi_{II}$ between the two platinum conductors I and II is easily calculated from eqs. (9-10) and (9-11), under the previously discussed conditions, by integration over all phases and phase boundaries. The result is:

$$E = \frac{1}{4F} \int_{\mu_{O_2}^{II}}^{\mu_{O_2}^{I}} t_{ion} \, d\mu_{O_2} \tag{9-13}$$

When $t_{ion} \simeq 1$, this equation reduces to equation (9-9) which was derived in another way [28, 46]. However, in contrast to eq. (9-9), eq. (9-13) is also applicable even when the electrolyte does not exhibit purely ionic conduction in the range of oxygen potentials being studied ($t_{ion} < 1$), and when t_{ion} varies locally across the electrolyte.

The preceding considerations also apply when electron holes rather than electrons are the minority defects in the electrolyte. Eq. (9-11) must then be reformulated, but the result in eq. (9-13) is the same.

The transport number of the ions (or of the electrons) in the electrolyte will vary because the concentration of free electronic charge carriers is a function of oxygen potential or partial pressure in analogy with eq. (4-7), while the ionic partial conductivity is fixed through the doping and remains approximately constant. Once again, these general considerations may be clearly illustrated by taking the solid electrolyte $ZrO_2(+MeO)$ as an example. Oxygen can react to a slight extent with the anion vacancies. There are a large number of these vacancies as a result of the doping, and so their concentration remains essentially constant. The reaction may be written as:

$$\frac{1}{2} O_2\,(g) + V_{O^{2-}}^{\bullet\bullet} + 2\,e' = O_{O^{2-}}^{2-} \tag{9-14}$$

Since the equilibrium concentration of electrons is very small, the laws of ideal dilute solutions may be applied. It then follows from eq. (9-14) that the concentration of electrons varies as $p_{O_2}^{-1/4}$.

The mobility of the electronic charge carriers is much greater than that of the ions. Therefore, despite the low concentration of electrons, when the oxygen partial pressure is very low the electronic partial conductivity will be of the order of the ionic partial conductivity. The following equation may then be written for the ionic transport number [29]:

$$t_{\text{ion}} = \frac{\sigma_{\text{ion}}}{\sigma_{\text{ion}} + \sigma_{e'}} = \left(1 + \frac{\sigma_{e'}}{\sigma_{\text{ion}}}\right)^{-1} = \left[1 + \left(\frac{p_{O_2}}{p_{(-)}}\right)^{-1/4}\right]^{-1} \tag{9-15}$$

The final bracketed expression in eq. (9-15) is obtained from the equilibrium condition of eq. (9-14), and from eq. (5-17) for $\sigma_{e'}$ and σ_{ion}. The constant $p_{(-)}$ contains the ionic partial conductivity, the electrochemical mobility of the electrons, and the equilibrium constant of eq. (9-14).

Since the electrons and the electron holes in the crystal are in equilibrium with one another, the concentration of electron holes will vary inversely as the concentration of electrons through the relationship $e' + h^{\bullet} = 0$. As a result, electron hole conduction will occur in the electrolyte when the oxygen potential is high. In this case, the term $\sigma_{h^{\bullet}}/\sigma_{\text{ion}} = (p_{O_2}/p_{(+)})^{+1/4}$ must be included in the bracketed expression of eq. (9-15). By substituting eq. (9-15) into eq. (9-13), we then obtain an expression for the emf of a solid state galvanic cell when the oxygen potential at the electrodes is different, and when mixed ionic and electronic conduction occurs in the electrolyte. This equation reads:

$$E = \frac{RT}{F}\left[\ln\frac{p_{(+)}^{1/4} + p_{O_2}^{\text{II}\,1/4}}{p_{(+)}^{1/4} + p_{O_2}^{\text{I}\,1/4}} + \ln\frac{p_{(-)}^{1/4} + p_{O_2}^{\text{I}\,1/4}}{p_{(-)}^{1/4} + p_{O_2}^{\text{II}\,1/4}}\right]$$

The calculations can easily be modified for other electrolytes. A further conclusion can be drawn from eq. (9-13): If the transport properties of the solid electrolyte and the thermodynamic parameters of the reference electrode are known, then the thermodynamic quantities ΔG, a_i, and p_i can be measured. On the other hand, if the thermodynamics of both electrode systems are known, then the transport properties of the electrolyte can be studied. For example, by a simple rearrangement of eq. (9-13) it follows that:

$$4F\left(\frac{\partial E}{\partial \mu_{O_2}^{\text{I}}}\right)_{\mu_{O_2}^{\text{II}}} = t_{\text{ion}} \quad \text{for} \quad \mu_{O_2} = \mu_{O_2}^{\text{I}} \tag{9-16}$$

The subscript $\mu_{O_2}^{\text{II}}$ indicates a constant oxygen partial pressure at the left electrode.

9.2.3. Applications of solid state galvanic cells

In this section we shall discuss some typical applications of solid state galvanic cells in the study of solid state reactions. The use of these cells to obtain thermodynamic data has already been discussed. For example, the emf of the galvanic cell [24]

$$\text{Pt/Fe, FeO/ZrO}_2\text{ (+MeO)/Ni, NiO/Pt}, \quad \Delta G_A^0 = -2E_A F \tag{A}$$

is a measure of the free energy change of the displacement reaction $NiO + Fe = FeO + Ni$, as one can easily see by adding up the partial reactions at the two-phase boundaries of the electrolyte for the passage of 1 mole (2 Faradays) of oxygen ions from right to left. In the same way, it may be shown that the emf of the cell

$$\mathrm{Pt/Ni,\ Cr_2O_3,\ NiCr_2O_4/ZrO_2\ (+MeO)/Ni,\ NiO/Pt}, \quad \Delta G_B^0 = -2\,E_B F \tag{B}$$

is a measure of the free energy of formation of the spinel according to the reaction $NiO + Cr_2O_3 = NiCr_2O_4$ [30]. Values of the free energies of formation of ternary compounds from binary reactants are most important for theoretical calculations of reaction rates of spinel formation, silicate formation, etc. according to eq. (6-25). A large number of ΔG_B^0 ($\Delta G_{AB_2O_4}^0$) values has been measured between 600 °C and 1 600 °C by means of galvanic cells [31]. Such ΔG^0 values for the formation of spinels lie mostly between – 5 and – 10 kcal per mole. The lower temperature limit for the experiments is a result of the increasing resistance of the electrolyte. As the temperature decreases, the resistance becomes of the order of that of the insulation, and approaches the resistance of the voltage measuring instrument. There are a number of comprehensive review articles in which a large variety of applications of solid state galvanic cells to the study of the thermodynamics of solid phases are discussed along with the attendant experimental details [8, 9].

The two following examples will show how galvanic cells can be used in the direct study of diffusion-controlled solid state reactions and phase boundary reactions. These examples are illustrative of a large number of possible applications in the field of solid state reaction kinetics. Emphasis here will be placed upon the principles of the methodology. Further examples can be found in the literature cited above.

The first example is concerned with the study of tarnishing processes [32]. By applying an external voltage U between the Pt leads of the galvanic cell:

$$\mathrm{Pt/Ag/AgI/Ag_2S\ (Pt)/NiS/Ni} \tag{C}$$

one can fix the sulphur activity in the Ag_2S at a predetermined value in accordance with eq. (9-9). In this cell, AgI acts as a silver ion conductor. When the external voltage is applied, an ionic current will flow in the electrolyte AgI, and a corresponding electronic current will flow in the external circuit until electrochemical equilibrium is established, at which time $UF = -EF = \Delta G = RT \ln a_{Ag}(Ag_2S)$. We note here than one can write $\Delta G = \mu_{Ag}^0 - \mu_{Ag}$ as the free energy of reaction for the transport of silver from its standard state (left electrode) into the Ag_2S (right electrode). By combining this expression with the equilibrium condition for the reaction $2\,Ag + S = Ag_2S$, we obtain the relationship:

$$2\,UF = \Delta G_{Ag_2S}^0 - RT \ln a_S (Ag_2S) \tag{9-17}$$

Thus, under potentiostatic conditions, we can fix the sulphur potential at the Ag_2S/NiS phase boundary by applying the proper potential U as given by eq. (9-17). There will then be a sulphur potential gradient (and consequently, also a nickel potential gradient) in the NiS layer between the Ag_2S/NiS and NiS/Ni phase boundaries. The processes which then take place in the NiS will be exactly the same as those which would occur during tarnishing in a sulphur atmosphere of given sulphur activity. That is, Ni^{2+} ions will migrate towards the Ag_2S/NiS phase boundary, and electron holes will migrate in the opposite direction (see Fig. 8-1). The Ni^{2+} ions will react with Ag_2S at the phase boundary according to the reaction $Ni^{2+} + Ag_2S = NiS + 2\,Ag^+$.

The Ag^+ ions will then be removed through the AgI electrolyte in order to maintain the sulphur activity at its potentiostatically fixed value. At the same time, a corresponding electric current flows through the external circuit. This current can easily be measured. In this way, a very precise determination of the tarnishing rate can be made (1 mA · 1 sec corresponds to $1/2 \cdot 10^{-8}$ moles Ni^{2+}!). The situation is schematically illustrated in Fig. 9-7.

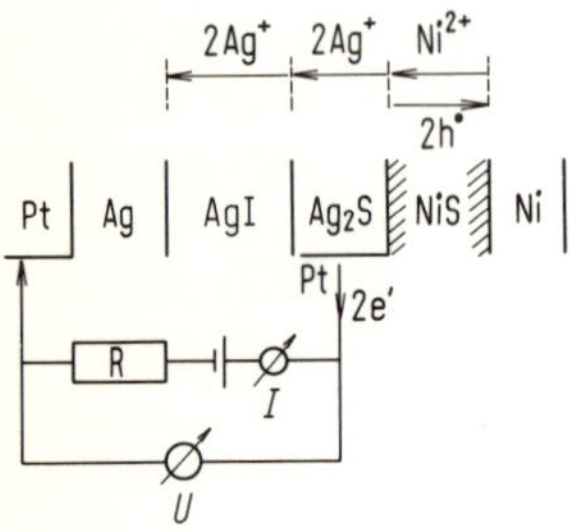

Fig. 9-7. Schematic diagram of a solid state galvanic cell to measure the tarnishing rate constant for NiS.

Conversely, by measuring the voltage under galvanostatic conditions, we can measure the instantaneous sulphur activity at the Ag_2S/NiS phase boundary for a given growth rate (current!) and a given thickness of the NiS product layer. It can be seen that all the essential parameters of tarnishing processes can be measured in a very simple and elegant way through the use of galvanic cells. A measurement of the current density I permits the parabolic reaction constant of eq. (8-6) to be calculated as follows:

$$\bar{k} = \frac{I V_{NiS} \Delta x_{NiS}}{2F} \qquad (9\text{-}18)$$

where Δx_{NiS} is given by:

$$\Delta x_{NiS} = V_{NiS} \int_{t=0}^{t} \frac{I}{2F} \, dt \qquad (9\text{-}19)$$

so that, finally:

$$\bar{k} = \frac{I(t) \int_0^t I dt}{4F^2} V_{NiS}^2 \qquad (9\text{-}20)$$

The activity gradient (chemical potential gradient) in the NiS can be determined by a measurement of the voltage U. Thus, by combining equations (8-8 a) and (9-20), the average component diffusion coefficient $\bar{D}_{Ni}$ of Ni in NiS can also be calculated. If the activity gradient in the NiS is not too large, then:

$$\bar{D}_{Ni} = \frac{RT}{4F^2} V_{NiS}^2 I(t) \int_0^t I dt \left[\frac{1}{\Delta G^0_{NiS} - \Delta G^0_{Ag_2S} + 2UF} \right] \qquad (9\text{-}21)$$

In the derivation of eq. (9-21), it should be noted that $\Delta G_{NiS} = \Delta G^0_{NiS} - RT\ln a_S$, and that $RT\ln a_S$ has been fixed equal to $\Delta G^0_{Ag_2S} - 2\,UF$ according to eq. (9-17), where U is the external voltage applied across the Pt leads of cell (C).

Let us now look at an example in which the rate of a phase boundary reaction is measured by an electrochemical method [33]. In order that only phase boundary effects are measured, it is necessary to ensure that all diffusion or convection processes are rapid in comparison to the phase boundary reaction. This condition may be achieved by proper choice of the dimensions of the sample, and, in the case of fluid phases, by forced or natural convection. The electrochemical method of measurement is particularly convenient and powerful, since the activities of the components as well as the reaction rates (as in the above example of the sulfidizing of Ni) can be directly measured electrically.

Once again, let us consider a cell in which Ag and Ag_2S are placed on either side of a solid AgI electrolyte. As shown in Fig. 9-8, the surface of the Ag_2S is in contact with a gas mixture H_2/H_2S. The sulphur activity is fixed by the H_2/H_2S ratio through the equilibrium condition for the reaction $H_2 + S = H_2S$.

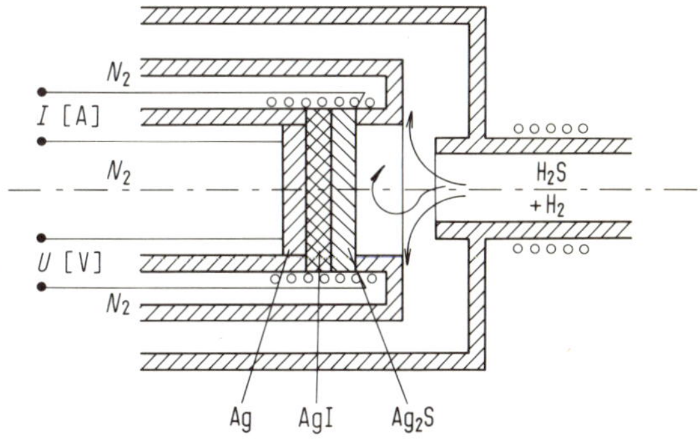

Fig. 9-8. Schematic diagram of the solid state cell $Ag/AgI/Ag_2S/H_2/H_2S$ for determining the rate of the phase boundary reaction during oxidation or reduction of Ag_2S with a H_2/H_2S gas mixture.

If an external voltage U is applied to the cell in Fig. 9-8 across the Pt leads, then the sulphur activity in the Ag_2S will be fixed by the magnitude of U (see eq. (9-17)). Either an oxidation or a reduction process involving the gas phase will then occur, depending upon whether the sulphur activity in the sulphide is greater than or less than it is in the surrounding gas phase. If the Ag_2S layer is thin enough (<1 mm at 300 °C), then the diffusional resistance can be neglected. Furthermore, under potentiostatic conditions, any change in the ratio Ag/S in the Ag_2S resulting from the oxidation or reduction processes with the gas phase will be compensated by a flux of Ag^+ ions through the AgI electrolyte. This flux of Ag^+ ions, which is accompanied by a corresponding measurable current in the external circuit, must occur in order that the sulphur activity in the Ag_2S remains at the value fixed by the voltage U. (It should be noted that Ag_2S is better written as $Ag_{2+\delta}S$ in order to clearly indicate the finite range of homogeneity.) Since the phase boundary reaction is rate-determining under the given conditions, the rate of the reaction:

$$H_2(g) + S^{n-}_{ad}\,(Ag_2S) = H_2S(g) + ne'(Ag_2S) \qquad (9\text{-}22)$$

can be directly measured in this way. The S^{n-}_{ad} are negatively charged sulphur ions which are adsorbed on the surface of the Ag_2S. They are assumed to be in equilibrium with the divalent sulphide ions and electrons in the interior of the sulphide. Through variation of the applied voltage U, the dependence of the reaction rate upon the component activities in the solid can be studied, as soon as steady state conditions are achieved. This dependence is a result of variations in the concentration of point defects in the bulk solid, which are in equilibrium with the defects in the surface. From such measurements, the reaction mechanisms at the surface can be studied. Information is obtained regarding the atomic species which take part in the phase boundary reaction. This goes far beyond the simple formulation of the overall reaction [33]:

$$H_2 + Ag_2S = H_2S + Ag(\text{dissolved in } Ag_2S)$$

The cell which is illustrated in Fig. 9-8 can also be used to measure the silver supersaturation which is required for the first nucleus of silver metal to form during the reduction of Ag_2S with H_2. First of all, the composition (given by δ) of $Ag_{2+\delta}S$ is fixed by application of an external voltage U across the cell under argon gas. Then the voltage source is disconnected. The silver sulphide is then exposed to hydrogen gas, and the potential of Ag in $Ag_{2+\delta}S$ is continually monitored by means of a high impedance voltmeter connected between the Ag and the Ag_2S (see Fig. 9-8). The result of such a measurement is shown in Fig. 9-9.

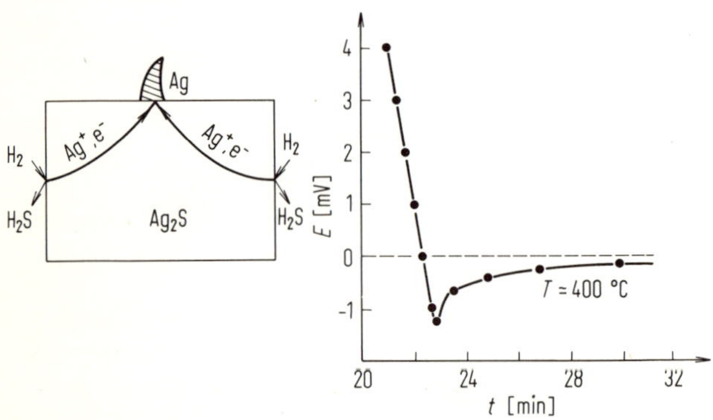

Fig. 9-9. Reaction rate and supersaturation during the reaction $Ag_{2+\delta}S + H_2(g) = (2+\delta)\,Ag + H_2S$, measured with a solid state galvanic cell of the type $Ag/AgI/Ag_{2+\delta}S/H_2(g)/Pt$.

Where the measured voltage passes through zero, the activity of Ag in the $Ag_{2+\delta}S$ is unity. Negative voltages are indicative of supersaturation ($a_{Ag} > 1$). It has been observed microscopically that the first silver nuclei form as needle-like whiskers at the point of maximum supersaturation [34].

These few examples of the application of solid state galvanic cells in the field of solid state reactions can only present a very limited view of this important area of solid state science. The examples were chosen primarily in order to demonstrate the principles according to which solid state research in thermodynamics and kinetics should be conducted with the use of electrochemical tools and methods. Such measurements are only possible because of the existence of suitable solid electrolytes. The most important of these are: $ZrO_2(+CaO)$ and $ThO_2(+Y_2O_3)$ for oxygen, silver halides and Ag_4RbI_5 for silver, copper halides for copper, some glasses in which certain ions are dissolved, and $\beta - Al_2O_3(+NaO)$ for sodium.

9.2.4. Fuel cells with solid electrolytes

The present interest in devices which will convert chemical energy directly into electrical energy is directed not so much towards the high energy efficiency of such devices as towards their high power density. The energy requirements of mobile units such as vehicles, rockets, and spaceships can only be met in this way [10, 35]. The key problem in the development of solid electrolyte fuel cells at present is the production of suitable electrodes. In Fig. 9-10 is shown a schematic diagram of a fuel cell based upon the oxy-hydrogen gas reaction:

$$H_2 + \frac{1}{2} O_2 = H_2O$$

The cathode reaction is:

$$\frac{1}{2} O_2 + 2\,e' = O^{2-}$$

and the anode reaction is:

$$H_2 + O^{2-} = H_2O + 2\,e'$$

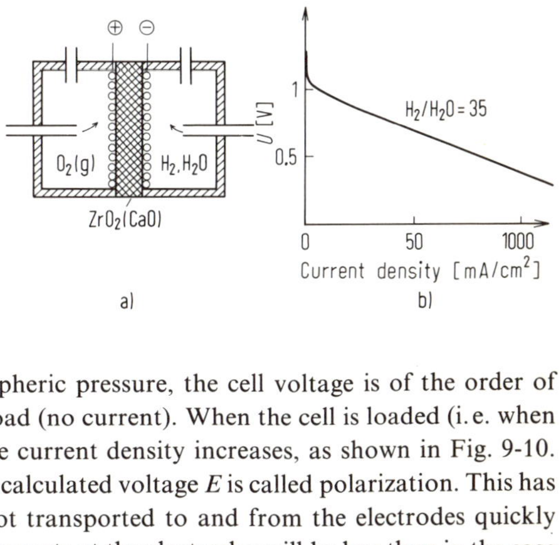

Fig. 9-10. (a) Schematic diagram of a fuel cell based upon the reaction $H_2 + 1/2\,O_2 = H_2O$. The current density versus voltage curve (b) is taken from the measurements of Binder et al. [36] at 1000 °C.

According to eq. (9-8), for gases at atmospheric pressure, the cell voltage is of the order of 1 volt at 1000 °C when the cell carries no load (no current). When the cell is loaded (i.e. when current flows), the cell voltage drops as the current density increases, as shown in Fig. 9-10. The deviation from the thermodynamically calculated voltage E is called polarization. This has essentially three causes: 1. If the fuel is not transported to and from the electrodes quickly enough, then the activities of the fuel components at the electrodes will be less than in the case of zero current. This is called concentration polarization. 2. If the actual charge transfer at the electrodes is the slowest step in the reaction sequence, then charge transfer polarization is said to occur. This is particularly prevalent at low temperatures. The key problem, which was mentioned above, lies in the development of suitable electrodes in which the activation energy of the charge transfer is reduced by catalytic action. 3. Finally, there is the ohmic resistance of the electrolyte and of the electrodes. This resistance polarization increases linearly with the current density.

The solid electrolyte fuel cells which are presently in the most advanced stages of development are those which employ ZrO_2-based electrolytes. Porous platinum or nickel are used as anodes, and platinum or silver are used as cathodes. The resistivity of the electrolytes is still of the order of 50 ohm cm even at 1 000 °C. Therefore, very thin plates or tubes of electrolyte must be used. As a consequence, problems of cracking and poor mechanical stability are encountered. However, it has been possible in practice to utilize hydrocarbon-steam mixtures as fuel with current densities of the order of 0.1 A/cm² at cell voltages of 0.5 volts.

9.3. Elementary processes of photography

9.3.1. Introductory remarks

Photography is a good example for the observation that a technology may frequently be far in advance of the scientific understanding of its elementary processes. For several decades, nearly everyone has been able to take photographs. However, the basic solid state reactions of the elementary photographic process are still the subject of lively discussion [4]. The actual light-sensitive emulsion consists of finely divided silver halide particles held in a gelatin base. The first clearly formulated working hypothesis of the mechanism of the elementary photographic process was presented in 1938 by Mott and Gurney on the basis of the defect theory of silver bromide. The most important aspects of this theory have already been presented in section 4.2.1. The essential results are as follows: 1. Silver halides possess a relatively high thermal intrinsic disorder of the Frenkel type (up to 1% just below the melting point of AgBr at 418 °C). The greater than exponential increase in conductivity in the neighbourhood of the melting point is at least partially due to the occurrence of anion vacancies (Schottky disorder). 2. In pure halides, electrons and electron holes occur as minority defects in very low concentrations. In thermodynamic equilibrium their concentrations are given as a function of the silver activity or of the partial pressure of the halide by eq. (4-24):

$$(h^{\bullet}) \propto (e')^{-1} \propto a_{Ag}^{-1} \propto p_{Br_2}^{\frac{1}{2}}$$

3. At room temperature, the electrical conductivity of AgBr is about 10^{-8} to 10^{-9} $\Omega^{-1}cm^{-1}$. The conductivity is ionic over almost the entire range of stability of the silver bromide. It is dependent upon doping by foreign ions, and is due to the motion of silver ions. The bromide ions are practically immobile. Only when the partial pressure of bromine is very high (i.e. only when the silver activity is of the order of 10^{-14} and less) does the electron hole conduction predominate. 4. At room temperature, the electrochemical mobility of the electrons is of the order of 50 cm²/V sec, while that of the electron holes is of the order of 1 cm²/V sec. Consequently, the deviation δ of silver bromide from the stoichiometric composition can be estimated with the help of eq. (5-17). Extremely small values of δ ($Ag_{1\pm\delta}Br$) are found. The silver deficit $\delta_{(-)}$ is formally calculated as 10^{-10}, while $\delta_{(+)}$ (silver excess) is even much less [19]. However, it should be noted that the effective mobilities of the electronic charge carriers are strongly dependent upon the existence of defect centers where they can be trapped for a relatively long time (trapping centers).

On the basis of these data for silver bromide, and by using the general transport theory for ionic crystals, we must now explain the following experimental fact: If AgBr crystals or a photographic emulsion are exposed to visible light of sufficient energy, then centers (also called the latent image) are formed which can be converted into visible silver particles through the use of a reducing agent. It has also been found that the photosensitivity of the silver halide is increased by plastic deformation (i.e. by the introduction of point defects and dislocations), and by doping with copper ions or with divalent anions such as S^{2-} or Se^{2-}.

What happens in the AgBr during the exposure to light? What is the nature of the centers which form the latent image? What takes place in the reducing solution (i.e. during developing)? In the following sections, these questions will be answered in terms of our present day understanding of the basic principles. It will be seen that the solid state reaction problem is very complex, but that it can be treated completely on the basis of the conceptual framework developed in the preceding chapters of this monograph.

9.3.2. The primary process of photography

The primary process of photography is the absorption of light. In terms of the band model of electrons, we say that, in the ideal crystal, an electron is promoted, with the absorption of light, from the valence band into the conduction band. Thus, the photon energy must be at least equal to the width of the energy gap between the two bands in order for absorption to take place. This energy is estimated in the literature to be 2.4 eV at $T=0$ K. This corresponds to a wave length of about 0.5 μ which lies in the visible region of the spectrum. If it is assumed that the valence band and the conduction band in AgBr are connected with the bromine and the silver respectively, then the absorption of light can be written in chemical language as:

$$h\nu + Br_{Br}^- + Ag_{Ag}^+ = Br_{Br} + Ag_{Ag} \qquad (9\text{-}23)$$

The partial reaction steps are:

$$Br_{Br}^- + h\nu = Br_{Br} + e' \quad \text{and} \quad Ag_{Ag}^+ + e' = Ag_{Ag}$$

where Br_{Br} and Ag_{Ag} are bromine atoms and silver atoms on regular lattice sites. That is to say, they symbolize electron holes in the valence band and electrons in the coduction band respectively. Now, the electronic defects are relatively mobile in the ideal crystal, and furthermore, they exert a coulombic attraction upon one another in the medium of dielectric constant ε. Therefore, it is to be expected that, after a short time, they will recombine by the reverse of reaction (9-23). The time required can be calculated by the methods of section 6.1.1. The energy which is set free can be emitted from the crystal as light or into the lattice as a phonon. However, the crystal is then left in the same state as before the absorption of light. No latent image or developable picture remains. We thus conclude that the photosensitivity of AgBr is a consequence of crystal defects. This conclusion is supported by the observation that the sensitivity is increased by dislocations and by foreign atoms. In any case, however, the primary process of photography involves the absorption of light with the creation of an electron-electron hole pair. Furthermore, it is observed that silver particles are the end product of the developing process. Therefore, the positively charged silver ions (which are quite mobile as a result of the Frenkel disorder) and the conduction electrons, which were created by the absorption of light, must react to form silver aggregates. In order for this to occur, however, the electrons and electron holes must not recombine with each other. That is, the electrons must combine with the silver ions and not with the electron holes, even though the electron holes are much more mobile than the silver ions. (From eq. (5-17) and the data in section 9.3.1, $u_{h^\bullet}/u_{Ag^+} \leqslant N_{h^\bullet}^{-1}$.) Therefore, a mechanism must be active whereby the electron holes are captured and rendered immobile in so-called traps before they can recombine with electrons. Possible traps for electron holes are bromine ions at dislocations. A center consisting of such an ion plus a trapped electron hole can then, in turn, exert a coulombic attraction upon silver ion vacancies (which bear an effective negative charge). In this way, complexes $[V'_{Ag}, Br^{\bullet}_{disloc}]^X$ are formed which correspond to a bromine atom on the dislocation. A similar situation can occur at outer surfaces. Copper (I) ions Cu^{+}_{Ag} can also act as traps for electron holes through the reaction [20]:

$$Cu^{+}_{Ag} + h^{\bullet} = Cu^{2+}_{Ag}$$

Furthermore, reaction of $h^{\bullet}$ with the substrate of the silver halide emulsion is also possible.

As a result of this trapping of electron holes, there will be enough time for the secondary process of photography to take place. The secondary process involves the reaction between conduction electrons and silver ions to form the silver nuclei of the latent image. Of course, it is conceivable – although rather improbable – that the reaction between the electrons and the silver ions is faster than the trapping of electron holes. In this case, the trapping process would be superfluous.

9.3.3. The secondary process of photography

When relatively mobile interstitial silver ions (which result from the Frenkel disorder) and conduction electrons from the primary process are available, then the secondary process can occur. This process consists of the reaction of these two species at crystallographically favourable sites [4, 5]. It is not possible for submicroscopic silver nuclei to form in undisturbed regions of the crystal. To be sure, neutral silver atoms can exist on lattice sites or in interstices, but metallic nuclei cannot originate here. However, these nuclei can form, for example, at jogs on dislocations. Such jogs possess a positive excess charge ($z_+ = \frac{1}{2}$). They can thus attract conduction electrons, thereby acquiring a negative excess charge. Positively charged interstitial ions will then be attracted to the sites and will precipitate there, so that the sites take on a net positive charge again. Electrons will now be attracted once again, and so on. This process will continue as long as photoelectrons from the primary process are available. Nucleation can also occur on a crystal surface, since the formation of nuclei is not spatially hindered there. Accordingly, we may speak of internal nucleation of the image upon dislocations in the interior of the crystal, or of external nucleation of the image upon an outer surface of the grain of silver halide. During the formation of the latent image in the form of submicroscopic silver nuclei, interstitial ions are constantly being used up. That is, the Frenkel disorder equilibrium is continually disturbed. Since the product $(Ag^{\bullet}_i) \cdot (V'_{Ag})$ remains constant at equilibrium according to eq. (4-19), new interstitial ions $Ag^{\bullet}_i$ and silver ion vacancies V'_{Ag} are continually supplied by means of the Frenkel defect formation reaction: $Ag_{Ag} = V'_{Ag} + Ag^{\bullet}_i$. This may happen as a homogeneous reaction in the bulk crystal, or, at these low temperatures, it may occur at inner and outer surfaces and dislocations, depending upon the kinetics, which will be most favourable for that process with the lowest activation energy. In section 6.1.1, it was shown that the relaxation time of the Frenkel defect formation reaction is very short, since this is a homogeneous reaction. Even at room temperature, the relaxation time was calculated to be of the order of milliseconds. Thus, equal amounts of vacancies and interstitial ions are formed practically immediately as the interstitial ions are used up to form the silver nuclei. Because of this fact, and because of the Frenkel equilibrium condition, the number of silver ion vacancies continually increases. Since these vacancies bear an effective negative charge, they can interact with the trapped electron holes to form associates or fixed defect complexes.

It should be mentioned in passing that, for silver halide grains with very small volumes, it is no longer meaningful to speak of concentrations in the sense of average values over (large) regions of space. However, it is conceptually valid to speak of time averages, or of averages over a large number of particles.

Finally, the role of sulphur in the secondary process of photography should be mentioned. It is now known that sulphur impurities in the gelatin substrate render the silver halide grains sensitive to light. The reason is as follows: At sites of particular instability in the silver halide – such as points where dislocations penetrate the surface – sulphur compounds react to form Ag_2S dissolved in the AgBr. The S^{2-} ions occupy anion sites. Therefore, as a result of the internal equilibria, both the concentration of interstitial silver ions $Ag_i^{\bullet}$ and the concentration of anion vacancies $V_{Br}^{\bullet}$ are increased. Furthermore, since S_{Br}' possesses a negative excess charge, it can act as a trap for the electron holes which were formed during the primary process. In this way the secondary process is aided. The overall sensitizing effect of the sulphur can be described by the overall reaction:

$$Ag_2S \xrightarrow{h\nu} AgS + Ag$$

or (9-24)

$$S_{Br}' \xrightarrow{h\nu} S_{Br}^{x} + e'$$

In the same way, the sensitizing effect of copper can be formulated by the following overall photochemical reaction:

$$CuBr + AgBr \xrightarrow{h\nu} CuBr_2 + Ag \qquad (9\text{-}25)$$

9.3.4. Developing the latent image

The latent image, which is formed by the action of light upon the silver halide, consists of submicroscope metallic aggregates of silver Ag_n ($n=2, 3, \ldots$). Overall electroneutrality is maintained by the existence of trapped electron holes. Now, as a result of thermal fluctuations, electron holes will, from time to time, break free from the trapping centers and will migrate in the valence band. These free electron holes can oxidize the silver nuclei. The silver ions so formed then become interstitial silver ions. The same result occurs when electrons break free from a silver nucleus and recombine with the electron holes. In this way, the effect of the light is reversed, and, after a finite time, the latent image can completely disappear. This dissolution of the latent image can be accelerated by increased temperature and by infra-red radiation. In either case, the energy which is supplied increases the rate at which electron holes are freed from traps or at which electrons are freed from the nuclei.

The silver nuclei can also be oxidized if the silver halide is in contact with an external redox system which acts to increase the concentration of electron holes. This process, which is the third important step in the photographic process, is the inverse of the so-called developing process which we shall now discuss.

When a photograph is developed, the submicroscopic particles grow into visible particles through the action of a reducing treatment. The overall process can be written as:

$$Ag_n + Ag_i^{\bullet} + e' = Ag_{n+1} \tag{9-26}$$

If $n = 1$, then the nucleus cannot be developed. If $n = 2$, development is possible after a long induction period. Nuclei can be developed normally with $n = 3$ or 4. The electrons in eq. (9-26) are supplied by an external redox system which is in contact with the AgBr. Eq. (9-26) may then be more completely formulated as:

$$Ag_n + Ag_i^{\bullet} + R^{r+} = Ag_{n+1} + R^{(r+1)+} \tag{9-27}$$

R^{r+} and $R^{(r+1)+}$ are the reduced and the oxidized forms of the developer. This process disturbs the Frenkel defect equilibrium, and thus interstitial ions and cation vacancies must continually be supplied as discussed in section 9.3.3. Therefore, for a given redox potential, the reaction (9-27) must eventually reach equilibrium for a certain value of n, unless the halide can dissolve in the coexisting redox phase. This situation can be formulated by adding the defect reaction equation:

$$AgBr = Ag_i^{\bullet} + V'_{Ag} + V_{Br}^{\bullet} + Br^-_{(solution)} \tag{9-28}$$

to eq. (9-27) to give the reaction for the developing process:

$$Ag_n + AgBr + R^{r+} = Ag_{n+1} + R^{(r+1)+} + Br^-_{(solution)} \tag{9-29}$$

The sum $V'_{Ag} + V_{Br}^{\bullet}$ corresponds to the removal of one lattice molecule, and so this sum drops out of the balance. Furthermore, we note that the reducing agent is generally of the form $H_2R^{r+}(aq)$, and so the developing reaction can be somewhat more realistically written as:

$$Ag_n + 2\,AgBr + H_2R^{r+}(aq) = Ag_{n+2} + R^{r+} + 2\,H^+(aq) + 2\,Br^-(aq) \tag{9-29a}$$

The redox potential, the potential of the halide in the solution, and the activity of silver in the AgBr are all coupled to one another through the equilibrium: $Ag + Br^-_{(solution)} = AgBr + e'$. The redox potential and the halide potential thus determine the silver activity, and so all defect concentrations are fixed at the phase boundaries [47]. All local transport coefficients are, in turn, determined by the defect concentrations. Therefore, all the prerequisites necessary for a quantitative treatment of the developing process are satisfied, provided that the spatial distribution and size distribution of the nuclei of the latent image are known, or that plausible assumptions regarding these distributions can be made, and provided that electrical double layer and space charge effects can be neglected.

However, Matejec [21, 22] has shown that electrical double layer and space charge effects should be included in a detailed discussion of the elementary processes of photography.

Since there is an excess of halide in the solution phase of a photographic emulsion, resulting in a depletion of $Ag_i^{\cdot}$ close to the surface of the crystal, the formation of the latent image near the surface is impeded relative to the interior of the crystal, as has been discussed in section 9.3.2. However, in the adsorption layer of the crystals, very mobile silver ions are expected to exist, which can, at least in part, counteract the depletion in the crystal near the surface.

In conclusion, it should be pointed out that the relative positions of defect energies are partially determined by the surroundings of the silver bromide, both in the emulsion and also in the developing bath. This fact may be largely responsible for the difficulties encountered in the quantitative interpretation of photographic processes.

9.4. Reduction of iron ores

9.4.1. General remarks

Only about one fifth of all metals occur in nature as elements or as alloys – and then usually in very small amounts. Therefore, a great deal of human effort has been, and is, devoted to the production of metals by reduction of their chemical compounds. A great number of different reduction processes have been used for the various elements, always with a view to economic optimization. Every reduction process is, in principle, based upon the overall chemical reaction:

$$\mathrm{MeX}_m + n\mathrm{R} = \mathrm{Me} + \mathrm{R}_n\mathrm{X}_m, \quad \Delta G \tag{9-30}$$

where R is the reducing agent. In order that the reaction proceeds in the proper direction, reducing agents must be sought which lead to a negative free energy change ΔG for reaction (9-30). In many important cases, the melting points of MeX_m and Me are so high that they exist as solid phases during the actual reduction process. This is unfavourable from the standpoint of the reaction kinetics. On the other hand, if possible, R is usually present as a gas, in order to yield a high reaction rate. When solid phases are present, then the reaction proceeds via the elementary steps of solid state reactions. This is the subject of this book. In the following, we shall show how the basic concepts which we have been discussing can be applied to describe the reduction of iron ore. Economically, this is still the most important metal reduction process today. No attempt will be made here to describe the technological aspects of the process. Rather, we shall discuss the elementary kinetic steps and their consequences.

Before we formulate the kinetics, let us examine the basis of the transport processes in iron and its oxides. The compound wüstite, $Fe_{1-\delta}O$, always has a cation deficit (see Fig. 9-11). This is the result of vacancies in the cation sublattice of the NaCl structure of the wüstite. Because of the condition of electroneutrality, the effective double negative charges of the cation vacancies are compensated by an equivalent number of trivalent iron ions. Despite the high concentration of defects, it appears that the laws of dilute solutions apply to the defects in wüstite to a good approximation [37, 38]. The structural complications, as discussed in section 4.3.1, which arise as a result of the interactions between point defects with high concentrations need not be dealt with here. The transport number of the electron holes in wüstite is approximately one. Because of the high concentration of vacancies in the cationic sublattice, the mobility of the cations is orders of magnitude greater than that of the anions.

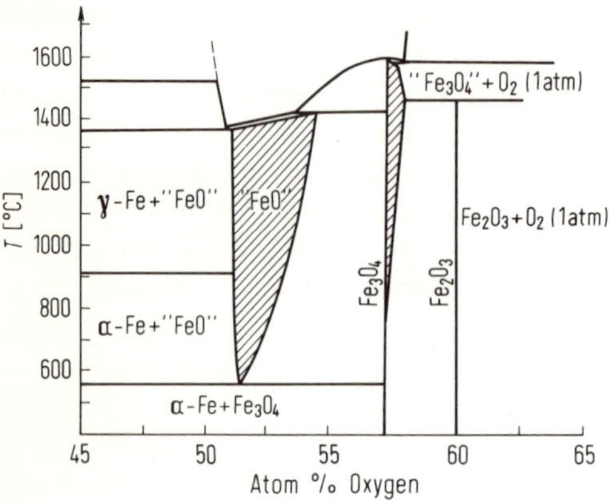

Fig. 9-11. Part of the Fe-O phase diagram.

Magnetite, $Fe_{3-\delta}O_4$, crystallizes in the spinel structure. From the phase diagram, it can be seen that this compound also has a deficit of cations. This can likewise be interpreted as resulting from cation vacancies and an equivalent number of electron holes (trivalent iron ions) [39]. The same comments regarding the relative mobilities of the ions and electron holes as were made for the case of wüstite also apply here.

Relatively little is known about the non-stoichiometry, disorder, and transport properties of hematite Fe_2O_3 (α-corundum structure). However, it can be said with certainty that Fe_2O_3 exhibits a small excess of cations, and that this excess is dependent upon the oxygen partial pressure. Furthermore, Fe_2O_3 is a n-type conductor over a wide range of oxygen partial pressures, and the diffusion coefficients of oxygen ions and of iron ions are of the same order ($\approx 10^{-11}$ cm^2/sec at $T = 1200$ °C) [40].

Oxygen transport through iron metal is also possible as a consequence of the solubility of oxygen in iron. The solubility in equilibrium with wüstite at 890 °C (α-iron) is about $5 \cdot 10^{-6}$ mole fraction, and at 1100 °C (γ-iron) it is about 10^{-4} mole fraction. The transport of oxygen through iron will be important if a compact surface layer of iron is formed during the reduction process. At the given temperatures, the diffusion coefficient of oxygen in iron lies between about 10^{-7} and 10^{-6} cm^2/sec.

9.4.2. Reduction of wüstite

First of all we shall describe the process of reduction of wüstite to iron [11]. We shall discuss the possible morphologies which may develop. Let us assume that our starting material is completely compact oxide with planar surfaces. The oxide is exposed to a reducing CO/CO_2 or H_2/H_2O gas atmosphere. The ratio CO/CO_2 is chosen so that the free energy change $\Delta G = \Delta G^0 + \sum RT \ln a_i$ for the reaction

$$FeO + CO = Fe + CO_2 \tag{9-30a}$$

is negative. The standard free energy change ΔG^0 for reaction (9-30a) can be obtained from appropriate tabulations [12]. CO molecules are adsorbed on the surface of the FeO. The surface concentration will depend upon the experimental conditions (P, T). These molecules then react with the oxygen at the wüstite surface. This oxygen can exist either as atoms or as ions and electron holes. The CO_2 which is formed is then desorbed. As a result, there will be concentration gradients of iron ions from the surface to the interior, and of electron holes from the interior to the surface of the wüstite phase. These gradients will, in turn, lead to diffusion. The overall result will be an increase in the average activity of iron in the wüstite. This activity will first achieve a value of one at the surface, at which time the Fe will be in equilibrium with wüstite of composition $Fe_{1-\delta^*}O$. The value of δ^* can be taken from the phase diagram for any particular temperature [41] (see Fig. 9-11). Before the first nucleus of iron forms, however, the activity a_{Fe} must actually become greater than one in order to overcome the barriers to nucleation. The necessary supersaturation for nucleation is probably quite small, as is indicated in the experiment described in section 9.2.3 and Fig. 9-9.

Following nucleation (which has the result that the supersaturation is more or less decreased), the reaction can take one of several possible courses: 1. A completely protecting surface layer of iron is gradually formed on the wüstite by lateral growth of the nuclei. Thereafter, the reduction rate is determined by the transport of oxygen in the Fe layer and by the reactions at the phase boundaries Fe/gas and FeO/Fe. As the thickness of the Fe layer increases, the diffusional resistance will eventually predominate, and the reaction rate will be determined by the product $c_O D_O$ in the iron (cf. eq. (5-4)). Let us assume that the oxygen dissolved in the iron obeys Henry's law for ideal dilute solutions, and also that the oxygen concentration at the Fe/gas phase boundary is negligibly small compared to the oxygen concentration c_O at the FeO/Fe phase boundary. This latter condition can always be met by a sufficiently large CO/CO_2 ratio. From Fick's first law, $j_O = -D_O \dfrac{c_O}{\Delta x}$, the parabolic reaction rate constant $\bar{k}$ for the increase in layer thickness $\mathrm{d}\Delta x/\mathrm{d}t$ is then calculated as:

$$\bar{k} = (1 - \delta^*)\, V_{Fe} D_O c_O \tag{9-31}$$

2. After nucleation, a coherent and compact surface layer of Fe is not formed. The reducing gas then always has access to the three-phase boundary FeO/Fe/gas. As long as the layer of iron sponge is relatively thin, the diffusional resistance to the transport of gas to and from the reaction sites will be relatively unimportant, and the chemical phase boundary reaction will be rate-controlling. Consequently, a linear growth law will be observed, as long as the interface area does not change with time. Experience indicates that, in contrast to the very early stages of the reduction of wüstite in the absence of a coexisting iron phase, the phase boundary reaction proceeds more rapidly on the Fe than on the FeO. Thus, the reaction rate is increased in the immediate vicinity of the Fe/FeO phase boundary at the surface as a result of the surface transport of oxygen over short distances upon the iron.

As the porous iron layer grows during the course of the reaction, the diffusion of CO through cracks and pores to the reaction site, and the diffusion of CO_2 away from the reaction site, can become rate-controlling. The boundary conditions at both ends of the pores are uniquely fixed (by the gas phase at the outer end, and by the equilibrium with Fe and FeO at the inner end). Consequently, the reaction rate can once again be calculated if the effective diffusion coefficient of the gases in the porous layer is known. The effective gas diffusion coefficient in a porous layer is a derived quantity which is given formally as the flux through a unit area per unit concentration gradient. In practice, the effective gas diffusion coefficient in porous materials can at best only be estimated, since this quantity depends upon the porosity, the pore size, and the tortuosity factor of the pores, as well as upon the diffusion coefficients of both the CO and CO_2 particles or, when the pressure is very low (when the mean free path length is greater than the pore diameter), upon the Knudsen diffusion coefficients. Once again, a parabolic rate law is expected.

3. The two preceding paragraphs have dealt with the most important morphological structures which develop during iron ore reduction. A further complication, not yet mentioned, may arise for the case of an initially completely compact product layer. A CO/CO_2 gas mixture has a certain carbon activity which is defined by the equilibrium condition for the reaction $2CO = CO_2 + C$. The carbon will dissolve, at a given activity, according to its solubility in the growing iron layer ($\underline{C}(Fe)$), and will diffuse to the Fe/FeO phase boundary where a pressure will be built up due to the reaction of $\underline{C}(Fe)$ with the oxide or with dissolved oxygen to form CO and CO_2 molecules. This pressure is determined by the equilibrium condition for the reactions: $\underline{C}(Fe) + FeO = CO + Fe$, and $\underline{C}(Fe) + 2FeO = CO_2 + Fe$. Depending upon the activity and the diffusion coefficient of the carbon, this pressure may be capable of bursting the iron layer from the oxide. It is also conceivable that, in certain cases, the CO and CO_2 molecules gather in cracks and micropores and are therefore enclosed in the otherwise compact iron layer. Limiting cases can be easily calculated with the aid of the known equilibrium constants. If the iron coating bursts, then a porous iron product layer should result. In certain cases, this will consist of a series of thin compact layers which are loosely held together.

9.4.3. Reduction of magnetite

Depending upon the reducing power of the gas, magnetite can be reduced either to wüstite or to iron as end product. During a reduction to wüstite, a non-porous compact reaction layer has been observed [48]. Local thermodynamic equilibrium was not attained at the wüstite/gas phase boundary. This indicates that the phase boundary reaction which occurs there has a high reaction resistance.

In order to demonstrate the principles of the formation of multiphase non-porous reduction layers, we shall now discuss the reduction of magnetite to iron. Even though the limiting case of non-porous reduced layers is only observed in exceptional cases, this discussion will nevertheless serve to illustrate the essential points, such as the coupling of fluxes at the phase boundaries.

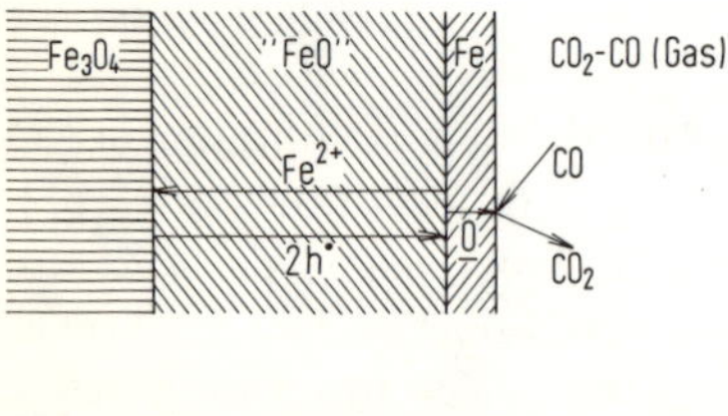

Fig. 9-12. Reaction scheme for the reduction of magnetite to metallic iron with the formation of a compact product.

The reduction experiment is schematically illustrated in Fig. 9-12. The overall reaction rate corresponds to the rate of diffusion of oxygen $\underline{O}$ through the Fe layer from the Fe/FeO phase boundary, where it is formed from the dissociation of FeO, to the Fe/gas phase boundary.

Part of the iron which is formed goes to build up the Fe layer. The rest diffuses in the form of cations through the wüstite via vacancies to the FeO/Fe_3O_4 phase boundary. Therefore:

$$\frac{1}{V_{Fe}} \frac{d\Delta x_{Fe}}{dt} = j_O(Fe) - j_{Fe}(FeO) \tag{9-32}$$

The increase in thickness of the wüstite layer is given by the number of moles of FeO which are formed per cm² at the FeO/Fe_3O_4 phase boundary according to the reaction:

$$Fe^{2+} + Fe_3O_4 = 4\,FeO + 2\,h^{\bullet}$$

minus the number of moles of FeO which dissociate at the Fe/FeO phase boundary. If $j_{Fe}(FeO)$ is the flux of iron ions through the wüstite to the FeO/Fe_3O_4 phase boundary, and if j_O is the flux of oxygen in the iron, then:

$$\frac{1}{V_{FeO}} \frac{d\Delta x_{FeO}}{dt} = 4\,j_{Fe}(FeO) - j_O(Fe) \tag{9-33}$$

Under the assumption of local equilibrium at the phase boundaries, the fluxes j_i in eqs. (9-32) and (9-33) can be written in the form $j_i = k^{(p)}/\Delta x^{(p)}$ for each product phase p. The reaction constant $k^{(p)}$ is discussed in section 6.2.3. It is not difficult to obtain a general solution of the coupled system of differential equations (9-32) and (9-33) if the substitution $\Delta x^{(p)} = \alpha^{(p)}\sqrt{t}$ is made. For each phase, the constant $\alpha^{(p)}$ is a function of the molar volume, the transport coefficients, and the driving forces.

We can obtain a better understanding of the physico-chemical process by simplifying the rate equations in the following way. We note that the oxygen solubility in iron is low, and thus the corresponding rate constant $\bar{k}$ (see eq. (9-31)) is relatively small. Consequently, for a quasi-steady state process, $\Delta x_{Fe} \ll \Delta x_{FeO}$, since the diffusional resistance of Fe per unit thickness is much greater than that of wüstite. It then follows from eqs. (9-32) and (9-33) that $j_O(Fe) \approx j_{Fe}(FeO)$. Eq. (9-33) can thus be written as:

$$\frac{d\Delta x_{FeO}}{dt} = 3\,k \frac{1}{\Delta x_{FeO}} \tag{9-34}$$

and this equation can be integrated directly.

The overall reaction rate $j_O(Fe) \approx j_{Fe}(FeO)$ can be calculated from the relation $j_{Fe} = \dfrac{k}{V_{FeO}\Delta x_{FeO}}$ (see above), provided that k is explicitly known. In order to calculate k, we proceed as follows: From eq. (5-13), the general flux equations may be written as:

$$j_{Fe} = j_{Fe^{2+}} = -\frac{c_{Fe^{2+}} \cdot D_{Fe^{2+}}}{RT} \cdot \frac{d\eta_{Fe^{2+}}}{dx} \tag{9-35}$$

$$j_{h^{\bullet}} = -\frac{c_{h^{\bullet}} D_{h^{\bullet}}}{RT} \cdot \frac{d\eta_{h^{\bullet}}}{dx} \tag{9-36}$$

where

$$\eta_{Fe^{2+}} = \mu_{Fe^{2+}} + 2\,F\phi \quad \text{and} \quad \eta_{h^{\bullet}} = \mu_{h^{\bullet}} + F\phi$$

Because of the condition of electroneutrality, $2\,j_{Fe^{2+}} = j_{h^{\bullet}}$. Furthermore:

$$c_{Fe^{2+}} D_{Fe^{2+}} \ll c_{h^{\bullet}} D_{h^{\bullet}}$$

since the transport number of the electronic charge carriers is approximately unity. It then follows from the reaction equation $Fe + 2\,h^{\bullet} = Fe^{2+}$ that, under conditions of local equilibrium, $\mu_{Fe} + 2\,\mu_{h^{\bullet}} = \mu_{Fe^{2+}}$. By making these substitutions, we arrive at the following expression for the flux of iron:

$$j_{Fe} = -\,c_{Fe^{2+}} D_{Fe^{2+}} \frac{d\ln a_{Fe}}{dx} \simeq -\frac{D_{Fe^{2+}}}{V_{FeO}} \frac{d\ln a_{Fe}}{dx} \tag{9-37}$$

Since the flux of iron ions is constant in the first approximation, we can integrate over the wüstite layer to obtain the equation:

$$j_{Fe} \simeq -\frac{1}{\Delta x_{FeO}} \int\limits_{a_{Fe}=1}^{a_{Fe(FeO/Fe_3O_4)}} \frac{D_{Fe^{2+}}}{V_{FeO}}\, d\ln a_{Fe} \tag{9-38}$$

Therefore, from eq. (9-34) and in agreement with eq. (8-9):

$$k \simeq \int\limits_{a_{Fe}=1}^{a_{Fe(FeO/Fe_3O_4)}} D_{Fe^{2+}} d\ln a_{Fe} \simeq \frac{\Delta G^0_{FeO}}{RT}\, \bar{D}_{Fe} \tag{9-39}$$

The parabolic rate constant for the growth of the wüstite layer during the reduction of magnetite under the conditions described above is then given from eq. (9-39) and eq. (9-34) as $\frac{3\Delta G^0}{RT}\,\bar{D}_{Fe}$. It is worthwhile to compare eq. (9-39) with eq. (8-8a) or eq. (6-23) in order to realize that the parabolic rate constant of eq. (9-39) is the analogue of the practical rate constants defined earlier.

From the above discussion, it is obvious how one would proceed in a quantitative treatment of the reduction of hematite, Fe_2O_3, in a suitable gas atmosphere. In this case, there are three coupled differential equations, and so the calculations are somewhat lengthier. Nevertheless, the solution follows in a straightforward manner from the substitution $\Delta x^{(p)} = \alpha^{(p)} \sqrt{t}$. However, it must be stressed here that the formation of a compact product layer has not been observed during the reduction of hematite, but rather, pores and cracks occur. Therefore, the calculations discussed above are not applicable.

The reduction of magnetite to iron with the formation of a porous metal layer can also be treated quantitatively in an analogous manner, as long as gas diffusion in the pores of the metal, and not a phase boundary reaction, is rate-determining. In this case, one needs only express the oxygen flux in eqs. (9-32) and (9-33) in terms of the appropriate effective diffusion coefficients for gas transport in porous layers.

9.4.4. Some practical problems

When we consider the practical problems of ore reduction, we realize how idealized the preceding discussion has been [11]. For one thing, especially at the beginning of the reduction, the resistance of the phase boundary reaction (for example, at the Fe/gas phase boundary) may surpass the diffusional resistance, and the phase boundary reaction will then be rate-controlling. In this case, the oxygen flux in eqs. (9-32) and (9-33) will be constant. If local equilibrium is maintained at the other phase boundaries, then the system of equations has a general closed solution. Furthermore, if $\Delta x_{Fe} \ll \Delta x_{FeO}$, then $j_O(Fe) \approx j_{Fe}(FeO) = \text{const}$, and the thickness of the product layer increases linearly with time.

In practice, there is a large number of complications which make a theoretical analysis of the reduction problem very difficult. We can only touch upon these problems here. There is, for instance, the question of nucleation, which is partially responsible for the morphology of the reduction product. The nucleation process will depend not only upon the degree of supersaturation, but also upon the condition of the surface and upon the impurity content of the ore. Another question is that of the porosity of the ore and of the reduction product which is formed upon the pieces of ore. The kinetics of the reduction will be strongly dependent upon any changes in the size and shape of the pores during the reaction. Finally, special problems arise during the reduction of mixed oxides, particularly when the oxygen affinities and the diffusion coefficients of the metals in the mixed oxide are very different. In order to give a better insight into the problems which can arise, let us look more closely at two of the complications mentioned above.

The influence of nucleation upon the reaction kinetics and the structure of the reaction product becomes especially important at low temperatures or when the reducing power of the gas is low. The supersaturation necessary for the formation of the first supercritical nucleus during the reduction of iron oxide is dependent upon many factors. This is hardly surprising if we consider that the rate of formation of critical nuclei (i.e. nuclei which can grow) is highly dependent upon the interfacial tension between the metal nucleus and the neighbouring phases. This was discussed in section 7.3. The interfacial tension γ_0 of a crystal is determined to a large extent by the content of dissolved impurities. The microroughness of the sample is also important in this regard. At high temperatures > 900 °C, it is observed that a supersaturation of the order of one percent is required, but at low temperatures, an iron activity in magnetite of 1.2 (i.e. a supersaturation of 20%) is possible [42, 43].

Microscopic investigations of the reduction of compact wüstite samples have shown that the condition of the surface and the porosity which are observed after the reduction are very strongly dependent upon the original concentration of defects in the wüstite. The greater is the initial concentration of vacancies, the rougher will be the surface, and the more pores will be found in the interior during the initial stage of reduction [44, 45]. It can be shown that planar interfaces are unstable under these given conditions. The concentration of cation vacancies at the surface is decreased because of the reduction reaction, while the vacancy concentration in the interior is still spatially constant. Therefore, the vacancy concentration gradient to convexities in the surface is less than the gradient to concavities. Hence, the flux of defects to the concavities is greater than that to the convexities. Consequently, the vacancy concentration at the concavities will be somewhat higher than at the convexities, as long as there is a finite phase boundary resistance to the removal of defects. Various authors have demonstrated that the reduction rate and the vacancy concentration are interdependent. An increased vacancy concentration results in an increased reduction rate. This fact, together with the increased concentration of vacancies at the concavities, leads to the aforementioned instability of the wüstite surface during reduction. Consequently, the greater is the original cation deficit of the wüstite sample, the greater will be the surface roughness.

In practice, the problem of the reduction of iron ores is extremely complicated, and in many respects, only a qualitative discussion is possible. Furthermore, in many cases the rate of reduction of iron ore in a blast furnace is controlled by heat transport, or by the rate at which material is supplied by streaming, since when the temperature is high enough, the diffusional processes and the phase boundary reactions will no longer be rate-determining. However, even in complicated situations, it is important to fully comprehend certain simplified cases, since it is through the study of such limiting cases that we achieve understanding and practice in the concepts of solid state reactions. With these conceptual tools, we are then ready to analyze more complicated situations, and to reach at least sound qualitative conclusions.

9.5. Literature

General Literature:

[1] A. L. Stuyts et al., Mater. Sci. Eng. *3*, 317 (1968/69).
[2] F. Thümmler and W. Thomma, Met. Rev. *115*, 69 (1967).
[3] H. Fischmeister and E. Exner, Metall *18*, 932 (1964), *19*, 113, 941 (1965).
[4] H. Frieser, G. Haase and E. Klein, Grundlagen der photographischen Prozesse mit Silberhalogeniden, Akademische Verlagsgesellschaft, Frankfurt 1968.
[5] O. Stasiw, Elektronen- und Ionenprozesse in Ionenkristallen, Springer-Verlag, Berlin 1959.
[6] F. v. Sturm, Elektrochemische Stromerzeugung, Chemische Taschenbücher, Verlag Chemie, Weinheim 1969.
[7] H. A. Liebhafsky, E. J. Cairns, Fuel Cells and Fuel Batteries, John Wiley and Sons, Inc., New York 1968.
[8] D. O. Raleigh in H. Reiss, Progress in Solid State Chemistry, Vol. 3, Pergamon Press, Oxford 1967.
[9] R. A. Rapp and D. A. Shores, Techniques of Metals Research IV, Part 2, Chapter VIC, Interscience Publ., New York 1970.
[10] K. Schwabe, Intern. Symp. Brennstoffelemente, Akademie-Verlag, Berlin 1968.
[11] L. v. Bogdandy and H.-J. Engell, Die Reduktion der Eisenerze, Verlag Stahleisen, Düsseldorf 1967.
[12] J. F. Elliott et al., Thermochemistry for Steelmaking, Addison-Wesley Publ. Comp., Reading 1963.

Special Literature:

[13] G. H. Jonker, Angew. Chem. *76*, 175 (1964).
[14] G. H. Jonker, Ber. Dtsch. Keram. Ges. *44*, 265 (1967).
[15] H. Salmang and H. Scholze, Die physikalischen und chemischen Grundlagen der Keramik, 5th ed., Springer-Verlag, Berlin 1968, p. 245.
[16] R. Shuttleworth, Proc. Phys. Soc. (London) *A 63*, 444 (1950).
[17] H. J. Oel and G. Tomandl, Das Sintern in der Keramik, Sonderdruck aus dem Inst. f. Werkstoffwissenschaften III der Universität Erlangen, February 1970.
[18] P. Reijnen in J. W. Mitchell, R. C. DeVries, R. W. Roberts and P. Cannon, Reactivity of Solids, Proc. 6. Intern. Symp. 1968, Wiley-Interscience, New York 1969, p. 93.
[19] C. Wagner, Ber. Bunsenges. phys. Chem. *63*, 1027 (1959).
[20] J. W. Mitchell, Rep. Progr. Phys. *20*, 433 (1957); J. Phys. Chem. *66*, 2359 (1962).
[21] R. Matejec and R. Meyer, Z. phys. Chem. NF *55*, 94 (1967).
[22] R. Meyer and R. Matejec, Z. phys. Chem. NF *59*, 251 (1968).
[23] L. S. Darken and R. W. Gurry, J. Amer. Chem. Soc. *67*, 1398 (1945).
[24] K. Kiukkola and C. Wagner, J. Electrochem. Soc. *104*, 379 (1957).
[25] R. Fitterer, J. Metals *19*, 92 (1967).
[26] F. Haber et al., Z. anorg. allg. Chem. *51*, 245, 289 (1906); Z. Elektrochem. *11*, 593 (1905), *12*, 415 (1906).
[27] E. Baur and H. Preis, Z. Elektrochem. *43*, 727 (1937).
[28] C. Wagner, Z. phys. Chem. *B 21*, 25 (1933).
[29] H. Schmalzried, Ber. Bunsenges. phys. Chem. *66*, 572 (1962).

[30] H. Schmalzried, Z. phys. Chem. NF *25*, 178 (1960).
[31] J. D. Tretjakow and H. Schmalzried, Ber. Bunsenges. phys. Chem. *69*, 396 (1965).
[32] S. Mrowec and H. Rickert, Z. phys. Chem. NF *36*, 329 (1963).
[33] P. Roy and H. Schmalzried, Ber. Bunsenges. phys. Chem. *71*, 200 (1967).
[34] C. Wagner and H. Schmalzried, Trans. AIME *227*, 539 (1963).
[35] W. Baukal, Chemie-Ingenieur-Technik *41*, 791 (1969).
[36] H. Binder et al., Electrochim. Acta *8*, 781 (1963).
[37] S. Takeuchi and K. Igaki, Sci. Rep. Tohoku Univ. *A 4*, 164 (1952).
[38] G. Lehmann, Ber. Bunsenges. phys. Chem. *73*, 349 (1969).
[39] H. Schmalzried and J. D. Tretjakow, Ber. Bunsenges. phys. Chem. *70*, 180 (1966).
[40] W. C. Hagel, Trans. AIME *236*, 179 (1966).
[41] M. Hansen: Constitution of Binary Alloys, 2nd ed., McGraw-Hill Book Comp., Inc., New York 1958.
[42] W. Morawietz and H. D. Schäfer, Arch. Eisenhüttenwes. *40*, 531 (1969).
[43] W. Morawietz and H. D. Schäfer, Arch. Eisenhüttenwes. *40*, 523 (1969).
[44] H. K. Kohl and H.-J. Engell, Arch. Eisenhüttenwes. *34*, 411 (1963).
[45] L. v. Bogdandy and H.-J. Engell, Die Reduktion der Eisenerze, Verlag Stahleisen, Düsseldorf 1967, p. 90.
[46] C. Wagner in P. Delahay, Advances in Electrochemistry and Electrochemical Engineering IV, Interscience Publ., New York 1966.
[47] E. Eisenmann and W. Jaenicke, Z. phys. Chem. NF *49*, 1 (1966).
[48] K. H. Ulrich, K. Bohnenkamp and H.-J. Engell, Arch. Eisenhüttenwes. *36*, 611 (1965).

List of symbols

(One often cannot avoid to use the same symbol for different terms. This list explains frequently used symbols. If symbols are given a different meaning, it is stated in the text.)

a_i	activity of component i ($=p_i/p_i^0$)
b_i	mobility of particles of sort i (velocity v_i per unit force)
c_i	concentration (number of moles of sort i per cm^3)
D_i	component diffusion coefficient of particles of sort i
D_i^*	tracer diffusion coefficient
$\tilde{D}_i$	partial diffusion coefficient
$\tilde{D}$	(chemical) interdiffusion coefficient
e_0	electron charge ($=1.6 \times 10^{-19}$ As)
e′	free electron in the crystal
E	electromotive force emf
f	correlation factor
F	Faraday constant ($=96487$ As equ.$^{-1}$)
G	Gibbs energy, free enthalpy
$\bar{G}_i$	partial molar Gibbs energy of component i ($\equiv$ chemical potential μ_i)
ΔG^0	standard value of the change in Gibbs energy for a reaction
h	Planck constant (6.626×10^{-27} erg s)
h˙	electron hole in the crystal
H	enthalpy (also $\bar{H}_i$, ΔH^0, see Gibbs energy G)
(i)	concentration of particles of sort i with respect to a lattice molecule
I	electrical current density
j_i	flux density of particles of sort i (mole cm^{-2} s^{-1})
k	Boltzmann constant ($=1.38 \times 10^{-16}$ erg K^{-1})
$\bar{k}$	practical (parabolic) reaction rate constant
k_j	reaction rate constant
m_i	mass of particle i
m^*	effective mass of an electron
n_i	mole number of particles of sort i
N_i	mole fraction of particles of sort i ($=n_i/\Sigma n_i$)
N_0	Loschmidt number (6.02×10^{23} per mole)
p_i	partial pressure of component i
P	pressure
R	universal gas constant ($=N_0 k = 8.32 \times 10^7$ erg mole^{-1} K^{-1} $=1.987$ cal mole^{-1} K^{-1})
S	entropy (also $\bar{S}_i$, ΔS^0, see Gibbs energy G)
t	time
t_i	transport number of particles of sort i ($=\sigma_i/\Sigma \sigma_i$)
T	absolute temperature (K)
u_i	electrochemical mobility of particles of sort i (velocity v_i per unit electrical field strength)
U	energy, internal energy
V	volume
V_m	molar volume
(V)	concentration of vacancies per lattice molecule

v_i	average velocity of particles of sort i
x	a coordinate in space
Δx	reaction layer thickness
z_i	electrical charge of ions of sort i
γ_i	activity coefficient ($=a_i/N_i$)
γ_0	surface free energy, surface tension
Γ_i	average jump frequency of particles of sort i
ε	relative dielectric constant
ε_0	absolute dielectric constant ($=8.854\times10^{-14}$ As V^{-1} cm^{-1})
η_i	electrochemical potential of particles of sort i ($=\mu_i+z_iF\phi$)
μ_i	chemical potential of particles of sort i $\left(\equiv\bar{G}_i=\left(\frac{\partial G}{\partial n_i}\right)_{P,T,n_j\neq n_i}\right)$
ν	vibrational frequency
ν_i	stoichiometric factor in a reaction equation (i.e. $\Sigma\,\nu_i A_i=0$)
ξ	special coordinate
ϱ	density (of mass, particles, charge, dislocations, etc.)
σ_i	electrical partial conductivity
σ	total electrical conductivity ($=\Sigma\,\sigma_i$)
τ	time interval, especially relaxation time
ϕ	electrical potential ($\mathfrak{E}=-\mathrm{grad}\,\phi=$ electrical field strength)

List of quantities, units (A), and physical constants (B)

(The units in this monograph are conventional. In order to facilitate a conversion into SI-units, the following listing is presented.)

(A)

Quantity	Unit	Symbol	Conversion
Length	Meter	m	1 m = 100 cm
Mass	Kilogram	kg	1 kg = 1000 g
Time	Second	s	
Temperature	Kelvin	K	
Electrical current	Ampere	A	
Force	Newton	N [$kg\,m\,s^{-2}$]	1 N = 10^5 dyn = 0.10197 kp
Pressure	Pascal	Pa [$N\,m^{-2}$]	1 Pa = 10^{-5} bar = 0.987×10^{-5} atm = 0.0075 Torr
Energy	Joule	J [N m ≡ W s]	1 J = 0.2388 cal = 2.78×10^{-7} kWh

(B)

Physical constant	Value and Unit	Symbol	Conventional Units
Speed of light	2.9979×10^8 [$m\,s^{-1}$]	c_1	2.9979×10^{10} [$cm^2\,s^{-1}$]
Electron charge	1.6022×10^{-19} [As]	e_0	4.8029×10^{-10} (esu)
Rest mass of electron	9.1096×10^{-31} [kg]	m_e	9.1096×10^{-28} [g]
Rest mass of proton	1.6726×10^{-27} [kg]	m_p	1.6726×10^{-24} [g]
Loschmidt number	6.0222×10^{23} [$mole^{-1}$]	N_0	
Gas constant	8.3143 [$JK^{-1}\,mole^{-1}$]	R	8.3143×10^7 [$erg\,K^{-1}\,mole^{-1}$]
Boltzmann constant	1.3806×10^{-23} [JK^{-1}]	k	1.3806×10^{-16} [$erg\,K^{-1}$]
Planck constant	6.6262×10^{-34} [J s]	h	6.6262×10^{-27} [erg s]
Faraday constant	9.6487×10^4 [As $equ.^{-1}$]	F	

Author index

Subject index

Table 1. Periodic table of elements.

Goldschmidt ionic radius is given for coordination number 6.

Atomic number, atomic weight, boiling temperature [°C], and melting temperature [°C].

	I A	II A	III T	IV T	V T	VI T	VII T	VIII T	
1	— — $^{1}_{1.01}$H 253 259								
2	1+ 0.78 $^{3}_{6.94}$Li 1330 109	2+ 0.34 $^{4}_{9.01}$Be 2770 1277							
3	1+ 0.98 $^{11}_{22.99}$Na 892 98	2+ 0.78 $^{12}_{24.31}$Mg 1107 650							
4	1+ 1.33 $^{19}_{39.10}$K 760 64	2+ 1.06 $^{20}_{40.08}$Ca 1440 838	3+ 0.83 $^{21}_{44.96}$Sc 2730 1539	4+ 0.64 (3+) (0.69) $^{22}_{47.90}$Ti 3260 1668	5+ 0.40 (3+) (0.65) $^{23}_{50.94}$V 3450 1900	6+ 0.35 (3+) (0.64) $^{24}_{52.00}$Cr 2665 1875	4+ 0.52 (3+) (0.70) $^{25}_{54.94}$Mn 2150 1245		
5	1+ 1.49 $^{37}_{85.47}$Rb 688 39	2+ 1.27 $^{38}_{87.62}$Sr 1380 768	3+ 1.06 $^{39}_{88.91}$Y 2927 1509	4+ 0.87 $^{40}_{91.22}$Zr 3580 1852	5+ 0.69 $^{41}_{92.91}$Nb 3300 2415	6+ 0.62 (4+) (0.68) $^{42}_{95.94}$Mo 5560 2610	7+ 0.55 ^{43} Tc — 2200	4+ 0.65 $^{44}_{101.07}$Ru 4900 2500	3+ 0.68 $^{45}_{102.91}$Rh 4500 1966
6	1+ 1.65 $^{55}_{132.91}$Cs 690 29			4+ 0.84 $^{72}_{178.49}$Hf 5400 2222	5+ 0.68 $^{73}_{180.95}$Ta 5425 2996	6+ 0.62 (4+) (0.68) $^{74}_{183.85}$W 5930 3410	6+ 0.55 (4+) (0.75) $^{75}_{186.20}$Re 5900 3180	4+ 0.67 $^{76}_{190.20}$Os 5500 2700	4+ 0.66 $^{77}_{192.20}$Ir 5300 2454
7	— — ^{87} Fr — 27								

			— — ^{61} Pm — 1027	3+ 1.02 (2+) (0.93) $^{62}_{150.35}$Sm 1900 1072
7	4+ 1.10 $^{90}_{232.04}$Th 3850 1750	2+ 0.93 ^{91} Pa — 1230	6+ 0.83 (4+) (1.05) $^{92}_{238.03}$U 3818 1132	4+ 1.04 ^{93} Np — 637

Key:

Ion valency	Radius
Atomic Number Atomic Weight	Element
Boiling Temperature Melting Temperature	